NAVIGATING

through

MEASUREMENT

in

GRADES 3–5

Nancy Canavan Anderson
M. Katherine Gavin
Judith Dailey
Walter Stone
Janice Vuolo

Gilbert J. Cuevas
Grades 3–5 Editor

Peggy A. House
Navigations Series Editor

WITHDRAWN

NATIONAL COUNCIL OF
TEACHERS OF MATHEMATICS

Copyright © 2005 by
The National Council of Teachers of Mathematics, Inc.
1906 Association Drive, Reston, VA 20191-1502
(703) 620-9840; (800) 235-7566; www.nctm.org

All rights reserved

Second printing 2007

Library of Congress Cataloging-in-Publication Data:

Navigating through measurement in grades 3–5 / Nancy Canavan Anderson ... [et al.] ;
Gilbert J. Cuevas, grades 3–5 editor.
        p. cm. — (Principles and standards for school mathematics navigations series)
    Includes bibliographical references.
    ISBN 13: 978-0-87353-544-8
    ISBN 10: 0-87353-544-8
      1. Mensuration—Study and teaching (Elementary) I. Anderson, Nancy Canavan. II.
Cuevas, Gilbert J. III. Series.
    QA465.N376 2004
    372.7--dc22
                                                                      2004020076

The National Council of Teachers of Mathematics is a public voice of mathematics education, providing vision, leadership, and professional development to support teachers in ensuring mathematics learning of the highest quality for all students.

Printed in the United States of America

TABLE OF CONTENTS

CONTENTS OF THE CD-ROM

Introduction

Table of Standards and Expectations, Measurement Pre-K–12

Applets

Stuck on Stickers

Geoboard

Blackline Masters and Templates

(All of those listed above, plus the following)

Guide to Constructing Picnic Items

Pattern Blocks

One-Inch Grid Paper

Geodot Paper

Two-Centimeter Grid Paper

Box Net 1

Box Net 2

Booth Labels

Award Ribbons

Readings from Publications of the National Council of Teachers of Mathematics

This Little Piggy

Daniel J. Brahier, Monica Kelly, and Jennifer Swihart

Teaching Children Mathematics

Water Matters

Becky Burghardt and Ginny Heilman

Teaching Children Mathematics

Mathematics Instruction Developed from a Garden Theme

Marta Civil and Leslie Khan

Teaching Children Mathematics

Exploring Measurement through Literature

Cheryl A. Lubinski and Diane Thiessen

Teaching Children Mathematics

Developing Spatial Sense through Area Measurement

Elizabeth Nitabach and Richard Lehrer

Teaching Children Mathematics

Third-Grade Students Engage in a Playground Measuring Activity

Anne Reynolds and Grayson H. Wheatley

Teaching Children Mathematics

About This Book

Navigating through Measurement in Grades 3–5 is the second of four books that demonstrate how teachers can introduce, develop, and extend some fundamental ideas about measurement at different grade levels. The introduction provides an overview of students' development of measurement concepts from prekindergarten through grade 12. Each chapter then addresses ideas of measurement that are basic in grades 3–5. Chapter 1 focuses on using informal or nonstandard units to teach the process of measurement, and chapter 2 examines students' understanding and use of standard units in the context of linear measures as well as measures of weight, angle, and temperature. Chapter 3 applies ideas about measuring length to the measurement of the area of two-dimensional objects, and chapter 4 extends these ideas to the volume, weight, and density of three-dimensional objects.

In each chapter, a discussion of key measurement ideas provides a foundation for activities that teachers can use to introduce these ideas to students and promote their familiarity with them. Each activity identifies the recommended grade level, the goals, the prerequisite skills and knowledge, and the necessary materials. Some of the activities include blackline masters for teachers to use directly with students. These activity pages, identified in the lists of materials and signaled by an icon in the margin, appear in the appendix. Teachers can also print them for classroom use from the CD-ROM that accompanies the book.

A second icon in the margin signals a variety of supplemental materials that appear on the CD. These include two computer applets for students to manipulate to extend their understanding of important concepts, as well as related readings for teachers' professional development.

All the activities share the same format. Each consists of three sections: "Engage," "Explore," and "Extend." "Engage" introduces tasks to capture students' attention and appeal to their interests. "Explore" presents the core investigation that all students should be able to do. "Extend" suggests additional activities for students who demonstrate continued interest and want to do some challenging mathematics. Margin notes include teaching tips, anticipated student responses to questions or activities, and references to some of the resources included on the CD-ROM.

Pertinent quotations from *Principles and Standards for School Mathematics* (National Council of Teachers of Mathematics [NCTM] 2000) also appear in the margin, signaled by a third icon. Assessment ideas for each activity include strategies for evaluating students, insights about students' performance, and ways to modify the activities for students who are experiencing difficulty or need enrichment. Finally, all activities provide ideas for connecting the concepts to related instructional topics.

As in the other books in the Navigations Series, the activities and ideas in this book are not intended to form a complete curriculum for measurement for this grade band. Rather, the authors hope that you will find in the book a collection of activities and investigations that you can use in conjunction with other instructional materials to enrich your curriculum and help your students develop a strong understanding of measurement.

Key to Icons

Principles and Standards

CD-ROM

Blackline Master

Three different icons appear in the book, as shown in the key. One alerts readers to material quoted from *Principles and Standards for School Mathematics,* another points them to supplementary materials on the CD-ROM that accompanies the book, and a third signals the blackline masters and indicates their locations in the appendix.

NAVIGATIONS **S**ERIES

GRADES 3–5

NAVIGATING *through* MEASUREMENT

Introduction

Measurement is one of the most fundamental of all mathematical processes, permeating not only all branches of mathematics but many kindred disciplines and everyday activities as well. It is an area of study that must begin early and continue to develop in depth and sophistication throughout all levels of learning.

In its most basic form, measurement is the assignment of a numerical value to an attribute or characteristic of an object. Familiar elementary examples of measurements include the lengths, weights, and temperatures of physical things. Some more advanced examples might include the volumes of sounds or the intensities of earthquakes. Whatever the context, measurement is indispensable to the study of number, geometry, statistics, and other branches of mathematics. It is an essential link between mathematics and science, art, social studies, and other disciplines, and it is pervasive in daily activities, from buying bananas or new carpet to charting the heights of growing children on the pantry doorframe or logging the gas consumption of the family automobile. Throughout the pre-K–12 mathematics curriculum, students need to develop an understanding of measurement concepts that increases in depth and breadth as the students progress. Moreover, they need to become proficient in using measurement tools and applying measurement techniques and formulas in a wide variety of situations.

Components of the Measurement Standard

Principles and Standards for School Mathematics (NCTM 2000) summarizes these requirements, calling for instructional programs from prekindergarten through grade 12 that will enable all students to—

- understand measurable attributes of objects and the units, systems, and processes of measurement; and
- apply appropriate techniques, tools, and formulas to determine measurements.

Understanding measurable attributes of objects and the units, systems, and processes of measurement

Measurable attributes are quantifiable characteristics of objects. Recognizing which attributes of physical objects are measurable is the starting point for studying measurement, and very young children begin their exploration of measurable attributes by looking at, touching, and comparing physical things directly. They might pick up two books to see which is heavier or lay two jump ropes side by side to see which is longer. Parents and teachers have numerous opportunities to help children develop and reinforce this fundamental understanding by asking them to pick out the smallest ball or the longest bat or to line up the teddy bears from shortest to tallest. As children develop an understanding of measurement concepts, they should simultaneously develop the vocabulary to describe them. In the early years, children should have experience with different measurable attributes, such as weight (exploring *heavier* and *lighter*, for example), temperature (*warmer* and *cooler*), or capacity (discerning the glass with the *most* milk, for instance), but the emphasis in the early grades should be on length and linear measurements.

As children measure length by direct comparison—placing two crayons side by side to see which is longer, for example—they learn that they must align the objects at one end. Later, they learn to measure objects by using various units, such as a row of paper clips laid end to end. They might compare each of several crayons to the row and use the results to decide which crayon is longest or shortest. Another time, they might use a row of jumbo paper clips to measure the same crayons, discovering in the process that the size of the measuring unit determines how many of those units they need. Their experiences also should lead them to discover that some units are more appropriate than others for a particular measurement task—that, for example, paper clips may be fine for measuring the lengths of crayons, but they are not practical for measuring the length of a classroom. As their experience with measuring things grows, students should be introduced to standard measuring units and tools, including rulers marked in inches or centimeters.

Children in prekindergarten through grade 2 should have similar hands-on experiences to lay a foundation for other measurement concepts. Such experiences should include using balance scales to compare the weights of objects; filling various containers with sand or water and transferring their contents to containers of different sizes and shapes to explore volume; and working with fundamental concepts of time and learning how time is measured in minutes, hours, days, and so forth—although actually learning to tell time may wait until the children are a bit older. By the end of grade 2, children should understand that the fundamental process of measurement is to identify a measurable attribute of an object, select a unit, compare that unit to the object, and report the number of units. In addition, they should have had ample opportunities to apply that process through hands-on activities involving both standard and nonstandard units, especially in measuring lengths.

As children move into grades 3–5, their understanding of measurement deepens and expands to include the measurement of other attributes, such as angle size and surface area. They learn that different kinds of units are needed to measure different attributes. They realize, for example, that measuring area requires a unit that can cover a surface, whereas measuring volume requires a unit that can fill a three-dimensional space. Again, they frequently begin to develop their understanding by using convenient nonstandard units, such as index cards for covering the surface of their desks and measuring the area. These investigations teach them that an important attribute of any unit of area is the capacity to cover the surface without gaps or overlaps. Thus, they learn that rectangular index cards can work well for measuring area, but circular objects, such as CDs, are not good choices. Eventually, the children also come to appreciate the value of standard units, and they learn to recognize and use such units as a square inch and square centimeter.

Instruction during grades 3–5 places more emphasis on developing familiarity with standard units in both customary (English) and metric systems, and students should develop mental images or benchmarks that allow them to compare measurements in the two systems. Although students at this level do not need to make precise conversions between customary and metric measurements, they should form ideas about relationships between units in the two systems, such as that one centimeter is a little shorter than half an inch, that one meter is a little longer than one yard or three feet, that one liter is a little more than one quart, and that one kilogram is a little more than two pounds. They should also develop an understanding of relationships within each system of measurement (such as that twelve inches equal one foot or that one gallon is equivalent to four quarts). In addition, they should learn that units within the metric system are related by factors of ten (e.g., one centimeter equals ten millimeters, and one meter equals one hundred centimeters or one thousand millimeters). Students should clearly understand that in reporting measurements it is essential to give the unit as well as the numerical value—to report, for example, "The length of my pencil is 19 centimeters" (or 19 cm)—not simply 19.

In these upper elementary grades, students should also encounter the notion of precision in measurement and come to recognize that all measurements are approximations. They should have opportunities to compare measurements of the same object made by different students, discussing possible reasons for the variations. They should also consider how the chosen unit affects the precision of measurements. For example, they might measure the length of a sheet of paper with both a ruler calibrated in millimeters and a ruler calibrated only in centimeters and compare the results, discovering that the first ruler allows for a more precise approximation than the second.

Moreover, they should gain experience in estimating measurements when direct comparisons are not possible—estimating, for instance, the area of an irregular shape, such as their handprint or footprint, by covering it with a transparent grid of squares, counting whole squares where possible and mentally combining partial squares to arrive at an estimate of the total area. In their discussions, they should consider how precise a measurement or estimate needs to be in different contexts.

Measurement experiences in grades 3–5 also should lead students to identify certain relationships that they can generalize to basic formulas. By using square grids to measure areas of rectangles, students might begin to see that they do not need to count every square but can instead determine the length and width of the rectangle and multiply those values. Measurement experiences should also help students recognize that the same object can have multiple measurable attributes. For example, they might measure the volume, surface area, side length, and weight of a wooden cube, expressing each measurement in the appropriate units. From the recognition that multiple attributes belong to the same object come questions about how those attributes might be related. If the side length of a cube were changed, for instance, what would be the effect on the cube's volume or its surface area? Similar questions arise in comparisons between various objects. Would two rectangles with equal perimeters necessarily have the same area? What about the converse? Would two rectangles with equal areas necessarily have the same perimeter? All these measurement lessons should help students appreciate how indispensable measurement is and how closely it is tied to number and operations, geometry, and the events of daily life.

Understanding of and proficiency with measurement should flourish in the middle grades, especially in conjunction with other parts of the mathematics curriculum. As students develop familiarity with decimal numeration and scientific notation and facility in computation with decimals, applications involving metric measurements provide a natural context for learning. As students develop proportional reasoning and learn to evaluate ratios, comparisons between measurements, such as the perimeters or areas of similar plane figures, become more meaningful. Their study of geometry requires students to measure angles as well as lengths, areas, and volumes and lets students see how measurements underlie classifications of geometric figures. For example, they identify triangles as acute, right, or obtuse by evaluating measurements of their angles or classify them as equilateral, isosceles, or scalene by comparing measurements of their sides. Proportional reasoning, geometry, and measurement converge when students create or analyze scale drawings or maps. Algebraic concepts of function that develop in the middle grades have applications in relationships such as that linking distance, velocity, and time. In science classes, students use both measurement and ratios to develop concepts such as density (the ratio of mass to volume) and to identify substances by determining their densities. Through experimentation, they discover that water freezes at 0° Celsius or 32° Fahrenheit and boils at 100° Celsius or 212° Fahrenheit, and from these data they can develop benchmarks for comparing the two scales. (For example, they can see that a ten-degree change in the Celsius temperature corresponds to an eighteen-degree change in the Fahrenheit temperature or that a forecast high temperature of 30° Celsius signals a hot day ahead.)

Middle-grades students should become proficient in converting from one unit to another within a system of measurement; they should know equivalences and convert easily among inches, feet, and yards or among seconds, minutes, hours, and days, for example. They should develop benchmarks for both customary and metric measurements that can serve as aids in estimating measurements of objects. For example, they might estimate the height of a professional basketball player as about two meters by using the approximate height of a standard doorframe as a benchmark for two meters, or they might use a right angle as a basis for approximating other angle measurements like 30, 45, or 60 degrees. Although students do more computations of measurements such as areas and volumes during the middle grades than in the earlier years, they still need frequent hands-on measurement experiences, such as tiling a surface with square tiles, making shapes on a geoboard, or building a prism with blocks or interlocking cubes, to solidify their understanding of measurement concepts and processes.

By the time students reach high school, they should be adept at using the measurement concepts, units, and instruments introduced in earlier years, and they should be well grounded in using rates, such as miles per hour or grams per cubic centimeter, to express measurements of related attributes. As they engage in measurement activities during grades 9–12, students are increasingly likely to encounter situations in which they can effectively employ powerful new technologies, such as calculator-based labs (CBLs), graphing calculators, and computers, to gather and display measurements. Such instruments can report measurements, often with impressive precision, but students do not always understand clearly what is measured or how the technology has made the measurement. How a measurement of distance is obtained when a tape measure is stretched between two points is obvious; it is not so obvious when an electronic instrument reflects a laser beam from a surface. Thus, students need a firm foundation both in measurement concepts and in how to interpret representations of measurements and data displayed on screens.

Also during the high school years, students encounter new, nonlinear scales for measurement, such as the logarithmic Richter scale used to report the intensity of earthquakes (a reading of 3 on the Richter scale signifies an earthquake with ten times the intensity of an earthquake with a Richter-scale measurement of 2). Especially in their science classes, students learn about derived units, such as the light-year (the distance that light travels in one year, moving at the rate of $3(10^8)$ meters per second, or about 186,000 miles per second) or the newton (N) (the unit of force required to give an acceleration of 1 m/sec^2 to a mass of 1 kilogram). Students also extend ideas of measurement to applications in statistics when they measure certain characteristics of a sample and use those data to estimate corresponding parameters of a population. Students preparing for a more advanced study of mathematics begin to consider smaller and smaller iterations—infinitesimals, limits, instantaneous rates of change, and other measurement concepts leading to the study of calculus.

Applying appropriate techniques, tools, and formulas to determine measurements

To learn measurement concepts, students must have hands-on experiences with concrete materials and exposure to various techniques, such

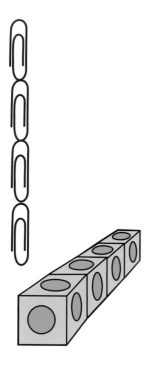

as counting, estimating, applying formulas, and using measurement tools, including rulers, protractors, scales, clocks or stopwatches, graduated cylinders, thermometers, and electronic measuring instruments.

In the pre-K–2 years, students begin to explore measurement with a variety of nonstandard as well as standard units to help them understand the importance of having a unit for comparison. Such investigations lead them to discoveries about how different units can yield different measurements for the same attribute and why it is important to select standard units. For young children, measurement concepts, skills, and the vocabulary to describe them develop simultaneously. For example, children might learn to measure length by comparing objects to "trains" made from small cubes, discovering as they work that the cubes must be placed side by side in a straight row with no gaps, that all the cubes must be the same size (though not necessarily the same color), and that one end of the object that they want to measure must be aligned with one end of the cube train. Later, when they learn to use rulers to measure length, they must learn how to locate the zero on the ruler's scale and align it with one end of the object that they are measuring. When they attempt tasks of greater difficulty, such as measuring an attribute with a unit or instrument that is smaller than the object being measured—the width of their desks with a 12-inch ruler or a large index card, for instance—they must learn how to iterate the unit by moving the ruler or card and positioning it properly, with no gaps or overlaps from the previous position. Furthermore, they must learn to focus on the number of units and not just the numerals printed on the ruler—counting units, for example, to determine that the card shown in the illustration is three inches wide, not six inches.

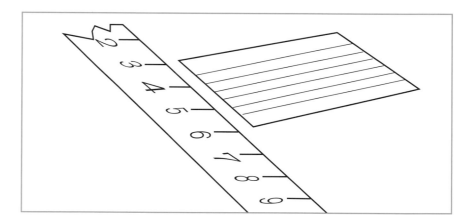

While students in prekindergarten through grade 2 are becoming acquainted with simple measuring tools and making comparisons and estimating measurements, students in grades 3–5 should be expanding their repertoires of measurement techniques and their skills in using measuring tools. In addition to becoming adept at using standard tools like rulers, protractors, scales, and clocks, third- through fifth-grade students should also encounter situations that require them to develop new techniques to accomplish measurement tasks that cannot be carried out directly with standard instruments. For example, to measure the circumference of a basketball, they might decide to wrap a string around the ball and then measure the length of the string; to measure the volume of a rock, they might submerge it in a graduated cylinder containing a known volume of water to obtain the total volume of water plus rock; to measure the weight of milk in a glass, they might weigh the empty glass as well as the glass and milk together.

As students in grades 3–5 hone their estimation skills, they should also be refining their sense of the sizes of standard units and the reasonableness of particular estimates. They might recognize 125 centimeters as a reasonable estimate for the height of a third grader but know that 125 meters or 1.25 centimeters could not be, or that a paper clip could weigh about a gram but not a kilogram. Students also should discuss estimation strategies with one another and compare the effectiveness of different approaches. In so doing, they should consider what degree of precision is required in a given situation and whether it would be better to overestimate or underestimate.

In grades 3–5, students also learn that certain measurements have special names, like *perimeter, circumference,* or *right angle;* and, as discussed earlier, they should look for patterns in measurements that will lead them to develop simple formulas, such as the formulas for the perimeter of a square, the area of a rectangle, or the volume of a cube. Through hands-on experience with objects, they should explore how different measurements might vary. For instance, by rearranging the seven tangram pieces to form a square, trapezoid, parallelogram, triangle, or nonsquare rectangle, they should find that the areas of all the shapes are the same, since they are made from the same seven pieces, but that the perimeters are different.

During middle school, students should apply their measurement skills in situations that are more complex, including problems that they can solve by decomposing or rearranging shapes. For example, they might find the area of an irregular shape on a geoboard by partitioning it into rectangles and right triangles (A) or by inscribing it in a rectangle and subtracting the areas of the surrounding shapes (B). Extending the strategy of decomposing, composing, or rearranging, students can arrive at other formulas, such as for the area of a parallelogram (C) by transforming it into a rectangle (D), or the formula for the area of a trapezoid either by decomposing it into a rectangle and two triangles (E) or by duplicating it to form a parallelogram with twice the area of the trapezoid (F). Other hands-on explorations that guide students in deriving formulas for the perimeter, area, and volume of various two- and three-dimensional shapes will ensure that these formulas are not just memorized symbols but are meaningful to them.

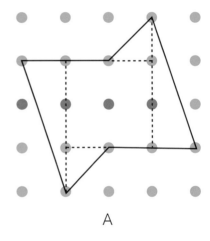

A

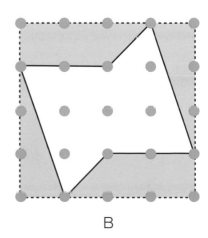

B

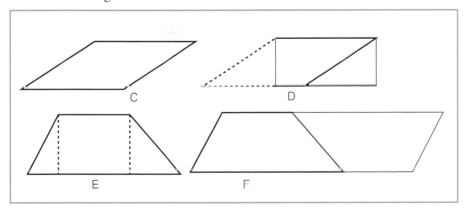

C D

E F

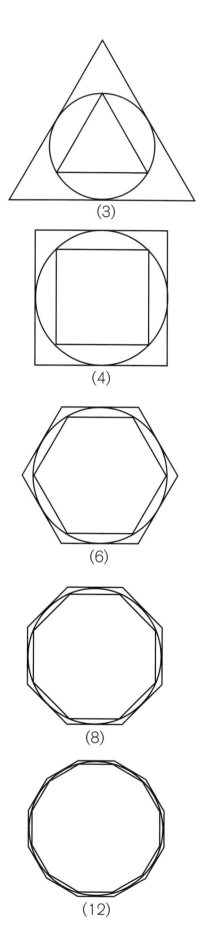

(3)

(4)

(6)

(8)

(12)

Students in grades 6–8 should become attentive to precision and error in measurement. They should understand that measurements are precise only to one-half of the smallest unit used in the measurement (for example, an angle measured with a protractor marked in degrees has a precision of ±0.5 degree, so a reported angle measurement of 52° indicates an angle between 51.5° and 52.5°). Students in the middle grades also spend a great deal of time studying ratio, proportion, and similarity—concepts that are closely tied to measurement. Students should conduct investigations of similar triangles to help them realize, for example, that corresponding angles have equal measures; that corresponding sides, altitudes, perimeters, and other linear attributes have a fixed ratio; and that the areas of the triangles have a ratio that is the square of the ratio of their corresponding sides. Likewise, in exploring similar three-dimensional shapes, students should measure and observe that corresponding sides have a constant ratio; that the surface areas are proportional to the square of the ratio of the sides; and that the volumes are proportional to the cube of the ratio of the sides.

Through investigation, students should discover how to manipulate certain measurements. For example, by holding the perimeter constant and constructing different rectangles, they should learn that the area of the rectangle will be greatest when the rectangle is a square. Conversely, by holding the area constant and constructing different rectangles, they should discover that the perimeter is smallest when the rectangle approaches a square. They can apply discoveries like these in constructing maps and scale drawings or models or in investigating how the shape of packaging, such as cracker or cereal boxes, affects the surface area and volume of the container. They also should compare measurements of attributes expressed as rates, such as unit pricing (e.g., dollars per pound or cents per minute), velocity (e.g., miles per hour [MPH] or revolutions per minute [rpm]), or density (e.g., grams per cubic centimeter). All these measurements require proportional reasoning, and they arise frequently in the middle-grades mathematics curriculum, in connection with such topics as the slopes of linear functions.

High school students should develop an even more sophisticated understanding of precision in measurement as well as critical judgment about the way in which measurements are reported, especially in the significant digits resulting from calculations. For example, if the side lengths of a cube were measured to the nearest millimeter and reported as 141 mm or 14.1 cm, then the actual side length lies between 14.05 cm and 14.15 cm, and the volume of the cube would correctly be said to be between 2773 cm³ and 2834 cm³, or (14.05 cm)³ and (14.15 cm)³. It would not be correct to report the volume as 2803.221 cm³—the numerical result of calculating (14.1 cm)³. Students in grades 9–12 also should develop a facility with units that will allow them to make necessary conversions among units, such as from feet to miles and hours to seconds in calculating a distance in miles (with the distance formula $d = v \cdot t$), when the velocity is reported in feet per second and the time is given in hours. Building on their earlier understanding that all measurements are approximations, high school students should also explore how some measurements can be estimated by a series of successively more accurate approximations. For example, finding the perimeter of

Navigating through Measurement in Grades 3–5

inscribed and circumscribed *n*-gons as *n* increases (n = 3, 4, 5, ...) leads to approximations for the circumference of a circle.

High school students can use their mathematical knowledge and skills in developing progressively more rigorous derivations of important measurement formulas and in using those formulas in solving problems, not only in their mathematics classes but in other subjects as well. Students in grades 9–12 should apply measurement strategies and formulas to a wider range of geometric shapes, including cylinders, cones, prisms, pyramids, and spheres, and to very large measurements, such as distances in astronomy, and extremely small measurements, such as the size of an atomic nucleus or the mass of an electron. Students should also encounter highly sophisticated measurement concepts dealing with a variety of physical, technological, and cultural phenomena, including the half-life of a radioactive element, the charge on an electron, the strength of a magnetic field, and the birthrate of a population.

Measurement across the Mathematics Curriculum

A curriculum that fosters the development of the measurement concepts and skills envisioned in *Principles and Standards* needs to be coherent, developmental, focused, and well articulated. Because measurement is pervasive in the entire mathematics curriculum, as well as in other subjects, it is often taught in conjunction with other topics rather than as a topic on its own. Teaching measurement involves offering students frequent hands-on experiences with concrete objects and measuring instruments, and teachers need to ensure that students develop strong conceptual foundations before moving too quickly to formulas and unit conversions.

The *Navigating through Measurement* books reflect a vision of how selected "big ideas" of measurement and important measurement skills develop over the pre-K–12 years, but they do not attempt to articulate a complete measurement curriculum. Teachers and students who use other books in the Navigations Series will encounter many of the concepts presented in the measurement books there as well, in other contexts, in connection with the Algebra, Number, Geometry, and Data Analysis and Probability Standards. Conversely, in the *Navigating through Measurement* books, as in the classroom, concepts related to this Standard are applied and reinforced across the other strands. The four *Navigating through Measurement* books are offered as guides to help educators set a course for successful implementation of the very important Measurement Standard.

NAVIGATING *through* MEASUREMENT

Chapter 1
Understanding the Process of Measurement

Principles and Standards for School Mathematics (NCTM 2000) emphasizes measurement as a content strand that provides opportunities for students to learn and apply "real world" mathematics. We use measurement ideas and skills throughout our lives. These concepts and capabilities enrich our learning of geometry, number, data analysis, and functions. The process that we use in measurement consists of three steps:

1. We determine the attribute that we want to measure. This might be the length of a path, the area of a rug, the temperature of water, or the weight of a stone, for example.

2. We select a unit that has the attribute that we wish to measure. For example, if we want to determine the attribute temperature in water in a container, the standard unit of measure is the degree (Fahrenheit or Celsius).

3. We compare the unit with the attribute of the object being measured. This involves some type of matching (lining up linear units alongside an object to determine its length, filling an object with cubic units to determine its volume, or covering an object with square units to determine its area, for example).

In prekindergarten through grade 2, students begin to develop an understanding of what it means to measure an object. They explore measurable attributes and use comparative terms, such as *heavier, lighter, longer, or shorter*. Before grade 3, students should have had a variety of experiences in measuring length, and they probably have participated in explorations of weight, volume, and time. In grades 3–5, students

Measuring with a nonstandard or informal unit of measure refers to using an available object as a temporary unit to measure an attribute of some other object. Using a pencil to measure the length of another object and using teddy bear counters to measure its weight are examples of such measuring.

need to acquire an understanding of the relationship between the objects that they are measuring and the units that they are using to measure them. Also, students should explore other attributes, such as area and angle, and ways to measure them. Students should use a variety of measurement tools and consider the accuracy of the measurements that these instruments give.

This chapter focuses on the development of third through fifth graders' understanding of the process of measurement through a series of explorations in which they use nonstandard or informal units to measure a variety of objects in familiar settings. The activities move from identifying measurable attributes of objects to creating and using nonstandard units of measurement in assorted everyday contexts. The first activity, What Can We Measure? gives students opportunities to explore measurable attributes of objects by finding common classroom items that they can use to measure these attributes in other commonplace objects, such as a piece of string to measure the length of a desk. In the second activity, Measurement Madness, the students compare objects on the basis of some measurable attribute, and they use nonstandard units as benchmarks for the comparisons. In Let's Measure, the third activity, the students consider some of the ways in which people have measured the attribute length, and they build on this investigation in creating an appropriate nonstandard unit to measure the attribute in classroom objects. The emphasis of this exploration is on helping students understand the relationship between a given attribute and the unit used to measure it. In the fourth activity, Ants' Picnic, the students pursue their exploration of nonstandard units, this time using them to measure paths around objects and thus explore the idea of perimeter.

What Can We Measure?

Grade 3

Goals

- Identify measurable attributes of a variety of classroom objects
- Create a list of measurable attributes of objects
- Select an object and identify how given attributes of the object can be measured with items found in the classroom

Prior Knowledge

Students should be familiar with attributes such as length, weight, volume, capacity, and area. They should understand weight as the measure of the pull or force of gravity on an object, mass as the amount of matter in an object and a measure of the force needed to accelerate it, and both volume and capacity as measures of the "size" of three-dimensional regions. (These descriptions are adapted from Van de Walle [2001], pp. 287–89.)

Materials and Equipment

- A collection of objects for each student in the class. These can be household items (soup cans and cereal boxes, for instance), classroom items (for example, books and crayon boxes), or other assorted items, such as wooden cubes, toys, and so on.
- Newsprint to construct a chart
- A copy of the blackline master "What Can We Measure?" for each student
- Overhead projector and markers

Classroom Environment

The students work together, first as a class, and then in small groups of four to five.

Activity

Engage

Discuss with the students the fact that all objects have measurable physical properties, such as weight, length, capacity, volume, or area. Show the children a classroom object—for example, a wastepaper basket. Ask, "Other than color, what attributes can we name to describe the basket?" Help students focus on the wastebasket's measurable attributes, such as its height (length), weight, and volume. Draw a picture of the basket on the board or an overhead projector. Label the different attributes that the students suggest as measurable. Have the students demonstrate the attributes that they identify. For example, if a student says that you can measure how tall the basket is, ask him or her to come up and show on the drawing what *tall* means.

Together with the class, create a chart like that in figure 1.1, listing the wastebasket under the heading "Object" and the characteristics that

p. 104

Volume and *capacity* both describe the space that three-dimensional objects occupy. If we are working with objects that contain liquids, we customarily express this space in units of capacity (pint, liter). Otherwise, we describe it as volume and express it in linear units that are cubed (cubic feet, cubic meters). In this book, the terms *volume* and *capacity* generally refer to the same concepts.

the class identifies under "Attributes That We Can Measure." You should reinforce the vocabulary of measurement by using the word height for "tallness," and you may need to prompt students to elicit some measurable attributes that they have not had previous opportunities to explore. For example, you can have a student hold the basket, and then you can ask, "Is it heavy or light?" After the student responds, you can ask the class, "What attribute are we identifying here?"

Object	Attributes that we can measure	How can we measure the attribute?
Wastebasket	The height (length) of the basket	Rods, toothpicks, straws, string
	The volume of the basket	Cubes, balls, cups of water
	The weight of the basket	Heavy-duty rubber bands that can stretch to form a spring scale

Turn next to the idea of using other objects in the classroom to measure these attributes. Point to the third column, "How Can We Measure the Attribute?" in the chart. Ask the students, "How could we measure, say, the height of the waste basket?" They will probably suggest a ruler or meterstick or other standard instrument. Ask them to think how they might make the measurement if they didn't have any of these tools. For example, what could they use to measure height if they couldn't leave the classroom and it contained no rulers or other standard measuring devices? Guide them in suggesting items in the classroom that they could use to measure each of the attributes of the waste basket.

Explore

With the students still seated at their desks, ask them to look around the room to identify an object that has some of the measurable attributes listed in the chart that the class has just created. Then arrange the students in groups of five, and ask one member from each group to get one such object to bring back to the team. Distribute copies of the blackline master "What Can We Measure?" to all students. Direct them to work individually to complete numbers 1 and 2, which call on them to write the name of the object selected and to draw a picture of it. Now the students are ready to work in their groups. Each student should describe to his or her teammates the measurable attributes of the object and identify other objects in the classroom that he or she could use to measure them. Observe closely to see if team members can suggest other ways to measure the attributes, and have them record all the suggestions in response to question 3 on the blackline master.

When the teams have created lists of different measurable attributes and ways to measure them, ask the students to share the objects and accompanying lists with the class. To summarize, record the students' findings on newsprint or an overhead transparency. Discuss with the students the informal or nonstandard units that they could use to measure the attributes.

Extend

Assign an attribute scavenger hunt as a homework activity. Give students a list of attributes—length, weight, volume/capacity, temperature, and area—and have them list an object for each of these attributes and suggest ways in which they could measure the attribute with nonstandard units. For example:

- Object—television screen
- Attribute—area
- Ways of measuring—index cards, squares of paper, tiles

Have the students report their results in a table similar to that in the blackline master "What Can We Measure?"

Assessment Ideas

You can use students' responses to the questions on the blackline master "What Can We Measure?" to determine what they understand about attributes and the different ways in which these can be measured. Give students a measurable attribute and ask them to look around the room, find an object that has that attribute, and demonstrate how they could measure it. For example, give a student the measurable attribute weight and ask her or him to find an object that could be measured by its weight. Then ask the student to show how she or he would measure the weight. The focus of the assessment should be on the student's understanding of the physical attributes of an object and nonstandard units that can be used to measure them.

Where to Go Next in Instruction

Students benefit from seeing how changing the nonstandard unit that they are using to measure an attribute can change the numbers in the measurement outcome. For example, how would the measurement of the length of an object change if the unit were doubled? What would happen if the unit used for measuring length were halved? Another important idea for teachers to introduce to students involves the appropriateness of the measurement unit. Would you use a single square tile to measure the area of the classroom floor, or a short length of string to measure the height of a door? Do these objects provide appropriate units for the measurement tasks? Students need to see a relationship between the magnitude of the measurement unit and size of the object whose attribute they intend to measure.

The next activity, Measurement Madness, explores the use of nonstandard units to estimate measurements. Explorations such as this one help students focus on the attribute being measured and the measurement process.

See "Third-Grade Students Engage in a Playground Measuring Activity" (Reynolds and Wheatley 1997) on the CD-ROM for useful ideas on extending measurement activities outside the classroom.

Measurement Madness

Grades 3–4

pp. 105–107

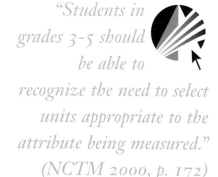

"All students
should …
understand such
attributes as length, area,
weight, [and] volume."
(NCTM 2000, p. 170)

"Students in
grades 3–5 should
be able to
recognize the need to select
units appropriate to the
attribute being measured."
(NCTM 2000, p. 172)

Goals

- Compare measures for the attributes length, weight, volume, capacity, area, and time
- Understand the process of measuring selected attributes with non-standard units
- Estimate measurements for selected attributes

Prior Knowledge

Students should have had experiences exploring length, area, perimeter, volume, capacity, weight, and time.

Materials and Equipment

- A copy of the blackline master "Measurement Madness" for each group of four students
- A permanent marker for the teacher's use in labeling objects

For the length station—
- A string about 7 inches long, for example

For the weight station—
- A set of 20 ten-penny nails held together with a rubber band, a set of 25 teddy bear counters, or 30 base-ten rods held together with a rubber band, for example
- A balance scale

For the volume station—
- A collection of 8–10 empty boxes of varying sizes, labeled A–J
- A benchmark unit of volume, such as a color cube or a unit constructed from other materials, such as sugar cubes or scoops of grains of rice

For the capacity station—
- A collection of 8–10 empty plastic jars of varying widths and heights, labeled A–J
- A container labeled as the benchmark unit of capacity
- A pitcher of water

For the area station—
- A collection of 10 rectangles of assorted, fairly similiar sizes, labeled A–J
- A benchmark unit of area, such as a 6-inch square, and a set of 250 color tiles (1-inch plastic squares) or 1-inch squares cut from construction paper

For the time station—

- A nonstandard unit of time, such as the quantity of time measured by an egg timer
- A chart listing the tasks specified for the time station

Classroom Environment

The classroom features six centers, or stations, devoted to length, weight, capacity, volume, area, and time. Students work in teams of four. Each student is assigned a number (1, 2, 3, or 4) to use as the team rotates through each center.

Activity

Engage

Explain to the students that they are going to work their way through six different measurement stations in teams of four. Each station is dedicated to a measurable attribute, and at each station, the students will find a nonstandard unit that they will use to measure that attribute in a number of objects. Some of the objects will be ones that they find around the classroom, and others will be objects that are waiting for them at a measurement station. Tell the students that they will have some special tasks to accomplish at the time station.

Ask the students to count off from 1 to 4 to form teams of four for the activity. Ask the students to remember their numbers. Give a copy of the blackline master "Measurement Madness" to each team, and point out that at each station, team members' numbers will determine who will act as the team recorder.

Use the weight station to model the activity. Say to students, "Let's suppose that a team has arrived at the weight station." Explain that the activity requires that each member of the team (except the recorder—here, student number 3) find an object in the classroom that is *lighter* than the set of nails (or whatever is serving as the nonstandard unit of weight), a second object that is of approximately equal weight, and a third object that is *heavier* than the nails. Each team member should heft the unit before he or she searches for the necessary items.

Have a team of students demonstrate. Each member should search the room and bring his or her three chosen objects back to the recorder at the weight station. The searchers should give statements about the weights of each of the objects that they have selected—for example, "I think this eraser weighs less than 20 ten-penny nails." The recorder should put the object on the balance scale and tell each team member whether she or he has made an accurate estimation.

Explore

During the activity, all the teams will rotate through all six stations, and students will be completing the tasks on the activity pages. The tasks that students are to perform at each station are as follows:

Length station. Team members will examine a length of string (or another object serving as a benchmark unit of length). Each will then find three objects—one that is longer than, another that is shorter than, and a third that is about the same length as the benchmark. The team recorder will verify the estimates by measuring the objects with the string.

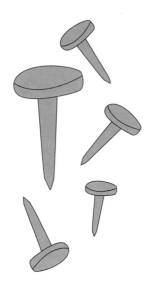

Weight station. Team members will heft a set of 20 ten-penny nails (or another weight serving as a benchmark unit). Each will then find something that is heavier than, something that is lighter than, and something that is about the same weight as the benchmark. The recorder will verify the estimates by putting the objects one at a time on the balance scale along with the benchmark.

Capacity station. Team members will examine the container labeled as the benchmark unit and the collection of jars labeled A–J. Each will then select a jar with a capacity that is less than, about the same as, or greater than the benchmark. The capacities of all the jars should be "close enough" to the benchmark unit that measurement is necessary to confirm the students' selections. The recorder will do this by pouring water from the pitcher into the benchmark unit and using it to fill each of the jars in turn.

Volume station. Team members will examine the box labeled as the benchmark unit and the empty boxes labeled A–J. Each will then select a box whose volume is less than, about the same as, or greater than the benchmark. Again, like the jars at the capacity station, all the boxes at the volume station should be "close enough" to the benchmark volume that measurement is necessary to confirm the students' selections. The recorder will do this by manipulating the benchmark unit in such a way as to use its components or contents to fill each of the boxes in turn.

Area station. Team members will examine the ten rectangles labeled A–J and the square designated as the benchmark. Each will then select a rectangle whose area is less than, about the same as, or greater than the benchmark. The recorder will verify the team members' selections by manipulating the benchmark unit to see if it could be made to "cover" each of the rectangles.

Time station. Team members will examine and experiment with the egg timer. On a chart at the station, students will find the following tasks:

A. Do 10 sit ups.
B. Walk around the classroom.
C. Sit down in your chair at your desk and stand up 5 times.
D. Write your full name neatly 10 times.
E. Sit down on the floor and stand up 5 times.
F. Draw a three-sided figure, a four-sided figure, a five-sided figure, and a six-sided figure, and write out your name.
G. List the birthdays of everyone on your team by month, day, and year.
H. List all the even numbers from 1 to 10.
I. List all the doubles from 1 to 5.

Students should take turns doing the tasks, timing themselves with the egg timer. The recorder should write the time that each task takes, and then the students should order the tasks from the longest to the shortest.

Extend

Encourage your students to extend their understanding of the connection between an attribute that they want to measure and the unit that they use to measure it. One way to do this is to have your students select attributes to measure in five classrooms items and ask the students to write a sentence explaining their choice in each case.

Assessment Ideas

Students' work at the measurement stations offers opportunities to assess both the recorders' and the other team members' understanding of measurement concepts. You can take note of the items that the team members select with attributes that are greater than, about the same as, or less than a given nonstandard unit for measuring those attributes. Have the students made appropriate estimations in the measurement comparisons? You can also observe the students who are acting as the recorders for their teammates. How appropriately have they judged a selected object? Do they give reasons for their evaluations? If so, what are the reasons? (For example, a student might say, "Look, I measured the area with the unit, and the rectangle you gave me doesn't have a smaller area than the unit.")

Where to Go Next in Instruction

Discuss the need for measurement instruments. Use examples from the activity that your students have just completed to support your argument. For instance, ask the students, "In the area station, where the unit was in the shape of a square, could you always use the unit easily and conveniently to measure the area of a rectangle?" Pursue the point: "Can anyone suggest something that might be easier, and explain why?" Students should note that using individual squares to cover the rectangles takes too much time. The next activity, Let's Measure, gives students an opportunity to construct a measurement tool and explore its uses. These experiences provide a basis for the introduction of conventional measuring tools and standard units of measurement.

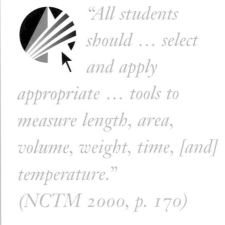

"All students should ... select and apply appropriate ... tools to measure length, area, volume, weight, time, [and] temperature." (NCTM 2000, p. 170)

Let's Measure

Grades 4–5

Goals

- Construct a measuring tool to measure length
- Use the measuring tool to measure objects in the classroom

Prior Knowledge

In prekindergarten through grade 2, students should have had experiences in measuring objects with nonstandard units and tools, such as tiles, paper clips, and pattern blocks, as well as with standard units, such as inches and ounces, and standard tools, such as rulers and scales. Students should have an understanding of measurable attributes, such as length, weight, volume, area, and temperature.

Materials and Equipment

For each pair of students—

- Two connecting cubes (about 5 cm long when connected)
- Two sheets of colored paper (different colors) that can be cut into strips
- A strip of construction paper (or oak tagboard), approximately 5 cm wide by 40 cm long
- A pair of scissors
- Paste or a glue stick
- A sheet of regular lined paper for a recording sheet
- Crayons, markers, or colored pencils

Classroom Environment

The students work in pairs.

Activity

Engage

Briefly explore the history of measurement with your students to identify different ways in which people have made measurements over time. Several Web sites provide relevant information (see the note in the margin). Provide an example, such as using the width of the palm of the hand as a measure of length. Explain that horse traders in the past measured horses' heights in hands, or palm widths, to avoid the inconvenience of carrying measuring sticks. Ask the students, "Why might units based on palms not be so good for measuring things?"

Show the reason with a model. Ask the students to consider the following hypothetical situation: "Let's say that our class needs a desk for a new student, and you are all going to measure your desks in palm widths so that it can be built." Have your students trace a hand with

Useful material about the history of measurement is available at numerous Web sites, including the following:

- http://cftech.com/BrainBank/ OTHERREFERENCE/ WEIGHTSandMEASURES/ MetricHistory.html

- http://www.slcc.edu/schools/ hum_sci/physics/tutor/2210/ measurements/history.html

- http://www.shaunf.dircon .co.uk/shaun/metrology/ feet.html

You may want to summarize the information to adjust its level for your students.

This activity has been adapted from Van de Walle (2001), p. 283.

their fingers extended and closed against one another and their thumb closed against the side of their hand. Then tell them to cut out the palm unit and use it to measure their desks in palms. Record the measures of the palm widths for several students on an overhead projector.

Ask the students if they would know how to build just one desk from the measurements that their classmates just recorded. Help them see that the measurements vary from person to person and that desks built with palm measures would be different sizes for most of the students in the class. You could illustrate the point with other palm cutouts that you have prepared beforehand, showing the palms of a baby, a kindergartner, an adult male, an adult female, and so on. Ask the students, "How could we use the information that we have collected on palm width to help ensure that two people could measure a desk independently with palms and come up with fairly consistent measurements?" Discuss with the students how using one palm width as a standard would help achieve consistency among measures.

Explore

Arrange your students in pairs, and distribute two connecting cubes and two sheets of differently colored paper to each pair. Explain to the students that they will be working in pairs to make a new style of ruler that they will then use to measure the lengths of objects in the classroom. Tell them to use the two cubes as their unit and to begin by measuring off lengths of paper that are equal to the unit. Distribute scissors and ask the students to cut out the lengths of paper. Give the pairs the longer strips of construction paper or tagboard, and tell them to paste their unit lengths to these longer strips in alternating colors, as shown in figure 1.2.

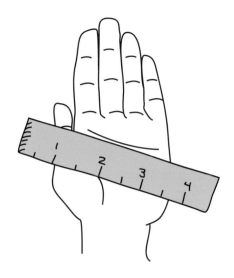

Fig. **1.2.**

Student-made ruler

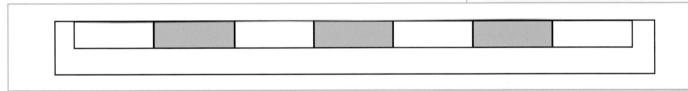

Let the students decide how many units they will include in their rulers. Encourage them not to use an edge of the strip as the starting point for the ruler, and explain that many conventional rulers are offset in this fashion.

Once the students have made their rulers, they will be ready to use them to measure the lengths of objects in the classroom. On a piece of lined paper, they should set up a recording sheet that they can use to list five to ten objects and their respective lengths as measured by their rulers. (See fig. 1.3.) Explain that they will all be measuring in the same units, with one unit equal to two cubes. Let the students practice measuring something with their rulers, and then have them label the units on the rulers, as shown in figure 1.4.

When the students have measured different objects with their calibrated rulers, have them post their lists of objects and the corresponding measurements. Help them look for patterns in the results by asking questions as such the following:

Fig. **1.3.**

Sample Recording Sheet

	Lengths Measured by Our Ruler	
	Objects	Lengths
1		
2		
3		
4		
5		
6		
7		
8		
9		
10		

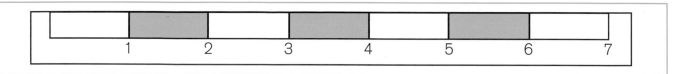

Fig. **1.4.**

Student-made ruler with the
units calibrated

- "Did the students who measured the same objects get the same measures?" (In theory, they should have.)
- "Why do you think there were differences in the lengths measured for the same objects?" (Differences may be due to faulty counting, imprecision in the construction of the rulers, or incorrect placement of the ruler while measuring the objects.

Extend

The preceding exploration offers an opportunity to introduce the idea of subunits and to discuss their importance in measuring accurately and precisely. Ask your students, "When you were measuring with your rulers, did you always measure the lengths of objects in whole units?" If the students gave measurements in fractions of a unit, how did they decide what fractions to use? Did they mark the subunits on their rulers? If they did, how did they make sure that they marked their subunits accurately? Did their subunits make sense? Were they reasonable? Were they easy to understand? Help the students think of ways to subdivide their units uniformly, and then ask them to remeasure several objects with the recalibrated rulers.

Students can also construct tools for measuring other attributes besides length. Possibilities include two-pan balance scales to measure mass and spring scales to measure weight. (See fig. 1.5.)

Fig. **1.5.**

Sample balance scale and spring scale

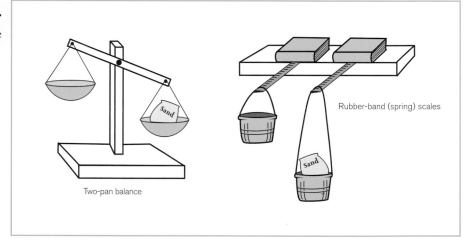

Rubber-band (spring) scales

Two-pan balance

Assessment Ideas

Constructing measurement tools facilitates students' explanations of the measures that they obtain. These explanations communicate a large amount of information about students' ideas of the measurement process and measurement units. Construction activities help students build a developmentally sound foundation for understanding standard units and using conventional measurement tools.

Where to Go Next in Instruction

Students should be given many measurement activities that use a variety of tools, such as rulers, balance scales, and thermometers. Instruction can progress from measuring with nonstandard units and student-created tools to measuring with standard units and conventional tools. Have students measure lengths of objects in the classroom with different standard tools, such as a 12-inch ruler, a meterstick, and a measuring tape capable of measuring several feet. For practice, use millimeters, centimeters, and meters as well as inches, feet, and yards. Have students record temperatures at particular times on successive days in both degrees Fahrenheit and degrees Celsius, and then have them graph the temperature changes over time. Grow a flowering plant, and ask the students to measure and record its growth and graph the data until the plant blooms. Students can use different types of scales to measure weight, such as a balance scale and a bathroom scale. Engage the students in determining the areas of regular as well as irregular shapes.

Activities that allow students to explore perimeter and area offer teachers ways to apply ideas related to the measurement process and the use of nonstandard units. Although area is treated in detail later in the book, the subject is previewed in the next activity, Ants' Picnic, which helps students develop the idea of perimeter while the shape that they are measuring is changing.

Ants' Picnic

Grades 3–4

Goals

- Use nonstandard units to explore the concept of perimeter
- Explore what happens to the measurement of a perimeter when the dimensions or size of the shape being measured changes

Prior Knowledge

The students should have had experiences measuring length.

Materials and Equipment

- A picnic basket (or something that can serve as one)
- A set of picnic items for each group of four students (see "Guide for Constructing Picnic Items" on the CD-ROM)
- 40–50 paper clips for each group
- A copy of the blackline master "Ants' Picnic" for each student
- A copy of *One Hundred Hungry Ants* by Elinor J. Pinczes (1993)

Classroom Environment

The students work in groups of four, but each child is responsible for his or her own work on the blackline master "Ants' Picnic."

Activity

Engage

Begin the lesson with all the cutouts of picnic items in a picnic basket or other container. Talk about going on picnics, and ask the students to think about what kinds of things they might expect to find in a picnic basket. Read aloud the story *One Hundred Hungry Ants* by Elinor J. Pinczes. Tell the students, "We are going to pretend to be a group of "picnic basket ants"—ants who make their way to picnics and into picnic baskets, like the ants in the story! But our special way of operating is to check out a picnic first by walking around the edge of the picnic area and then by walking around the edge of each of the picnic items. The way we work just might help us discover the meaning of a new term— *perimeter*."

On the board or an overhead transparency, draw a diagram of the classroom like that in figure 1.6. Use your drawing to help you model the exploration before the children begin work with the picnic basket items. (If your classroom is not rectangular, draw a different diagram that shows its shape.)

Select two or three students who have approximately the same shoe size. Have these students pretend to be the ants arriving at the picnic site—in this case, the classroom. Line the students up at the end of one of the classroom walls, and have them walk heel to toe, one student after another, to the opposite end of the wall, counting the number of

pp. 108–109

The "Guide for Constructing Picnic Items" on the CD-ROM tells how to use construction paper to make representations of a paper napkin, a placemat, a sandwich, and a brownie for each group of students to work with in Ants' Picnic.

One Hundred Hungry Ants, written by Elinor J. Pinczes and illustrated by Bonnie MacKain (1993), is a story in rhyme about one hundred ants marching to a picnic in single file. On the way, they decide that they could go faster in two lines of fifty, or even faster in four lines of twenty-five, or faster still in ten lines of ten. So much frantic reorganizing delays the ants, and they miss the picnic.

"All students should ... understand the need for measuring with standard units." (NCTM 2000, p. 170)

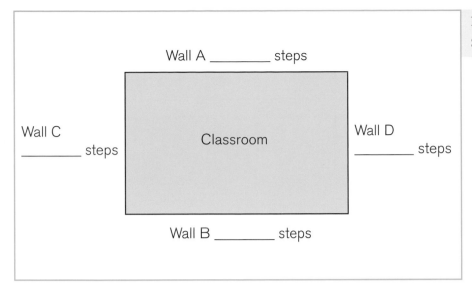

Fig. **1.6.**

Sample classroom diagram

Wall A _____ steps

Wall C _____ steps

Classroom

Wall D _____ steps

Wall B _____ steps

In "Exploring Measurement through Literature," available on the CD-ROM, Lubinski and Thiessen (1996) present ideas for using a variety of children's books in teaching measurement concepts.

steps that they take. Let them record the numbers as the measure of the length of wall A on the diagram. If their numbers are different, ask the class why one student took more or fewer steps than another. Discuss differences in results in relation to differences in shoe length, counting accuracy, or the students' precision in walking heel to toe.

If your classroom is rectangular, ask the students, "Can you think of a way to use the number of steps that we have recorded for wall A to label any other wall in the drawing?" (If the room is not rectangular, have the students walk heel to toe again to measure the length of each wall). Talk to the students about the fact that opposite walls should take the same number of steps, and ask them why this is so. (These walls are the same length if the room is truly rectangular.) Copy the number of steps from wall A onto the line labeled wall C on the diagram. Then ask the students to use the length of wall A or wall C to estimate how many steps the same students would take to walk wall B or wall D, shown as shorter on the diagram. To help them refine their estimate and come up with a number that they can agree on as a class, ask questions such as "Is Wall B more than half of wall A, or is it less than half?" and "How many times do you think wall B would fit on wall A?" Make sure the students are thinking about the classroom's walls instead of the walls shown on the diagram.

When all your students have settled on one estimate, record this number on the diagram. Then have the same two or three students return to the same corner of the classroom as before and walk heel to toe again, counting the number of steps that they take to walk the length of the wall adjacent to the first one. Record the results and compare them to the estimate. Then ask, "How can we find the number of steps for wall D?" (They might respond, "It is the same as wall B," or "We have to measure it by walking.")

Now your students should be ready to put together what they have learned to arrive at a new idea. Ask them, "How could you find the total number of steps that these students would need to take to go all the way around the classroom?" Relate this total number of steps to *perimeter*.

Explore

Distribute a copy of the blackline master "Ants' Picnic" to each student, and supply a set of the picnic items to each team of four students. These items include construction-paper representations of a placemat, a napkin, a sandwich, and a brownie. Give each team 40–50 paper clips, and have each team member take 10–12 of them, along with one picnic item. Let the students work through the tasks on the blackline master. Once the students have completed the activity, have each group share the definitions of *perimeter* that the group members have written. Record these on newsprint or an overhead transparency. Discuss any patterns that appear on the descriptions (for example, "The perimeter is the sum of the lengths of the sides," or "Perimeter is the total length around a shape.")

Extend

It is important for students to consider the perimeter of irregular two-dimensional figures. Draw an irregular shape on an overhead transparency and specify the length of each side. (See the example in fig. 1.7.) Ask students how they would determine the perimeter of the shape. (Add the measurements of the sides.)

Assessment Ideas

Invite the students to discuss their results. Listen to what they say to determine the extent to which they have understood the idea of perimeter.

Where to Go Next in Instruction

Ask a student to use paper clips to measure the perimeter of the classroom presented in the diagram at the beginning of the activity. Then ask the class, "What would happen to the numbers in the measure of the perimeter if we were to use a string that was longer than a paper clip to measure the sides?" (The perimeter would be a smaller number; you might show your students a length of string and let one of them use it to measure the perimeter of the classroom in the drawing.) Relate the differences in the measurements of the perimeter to the need for a uniform or standard unit of measurement that can provide a

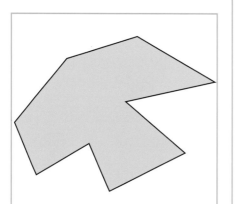

Fig. **1.7.**

Sample irregular shape

Students may be interested to know that peri *comes from Greek and means* around. *A* periscope, *for example, lets users look all around.* Meter *also has a Greek origin and means* measure. *Thus, students can see how the word* perimeter *was formed to mean the measure around a closed area.*

consistent measure of length. Ask the students to list any standard units of measure that they know about or have used before (e.g., inch, foot, meter).

Students need to be familiar with standard measurement units, and they should be able to use conventional measuring tools and standard units to make accurate measurements and reasonable estimates. The chapter that follows addresses these topics.

Opportunities to use informal or nonstandard units of measurement let students explore ideas related to the measurement of attributes and help them understand the reasons behind the development of standard units.

NAVIGATING *through* MEASUREMENT

Chapter 2
Understanding Standard Units of Measurement

Educators generally agree that students should explore and use informal or nonstandard units when they are beginning to develop measurement ideas and skills. Over time, however, students need to make a transition to standard units and conventional or standard measuring tools. Van de Walle (2001) identifies three broad goals that teachers should help students attain as they learn about standard measures. The students should develop—

1. a familiarity with standard units;
2. an ability to select an appropriate unit in a particular context; and
3. an understanding of relationships among units.

The activities in this chapter are designed to help students understand the need for measuring with standard units while making them familiar with some of these standard units in the customary (English) and metric systems. The first activity, Off to the Hardware Store, turns students' attention to standard units. In the second activity, Measurement Scavenger Hunt, students use standard units to measure length, weight, and angle.

It is also important for students in grades 3–5 to have experiences in estimating measures. Such experiences emphasize the process involved in measuring, and they familiarize students with measurement units. Students also need activities that help them develop and use appropriate language for estimating. When presenting estimates, students should make statements such as "This book is *about* 33 grams" and "The area of this carpet is *a little more than* 5 square yards," for example. Students will

29

explore the selection and use of benchmarks to estimate measurements in the chapter's third activity, My Benchmarks.

In addition, it is important for students in grades 3–5 to begin to understand that the size of the unit is related to the precision with which they can make a measurement. Moreover, students need to recognize that all measurements contain some error and that small units do not automatically guarantee precision in measurement. Students should begin to understand that the degree of precision that is necessary and appropriate in a particular measurement is related to the context and purpose of the measurement task at hand. An engineer measuring components of an automobile engine needs to achieve a higher level of precision than does a homeowner measuring carpet for a living room. The chapter's fourth activity, How Precise Should My Measurement Be? helps students understand the idea that all measurements are approximations and that differences in units affect precision.

Principles and Standards for School Mathematics (NCTM 2000) identifies the ability to make simple unit conversions within a system of measurement as one of the skills that students in grades 3–5 should master. Making such conversions helps students develop an understanding of relationships among units. In Conversion Sense, the last activity of the chapter, students carry out simple unit conversions within a system of measurement.

Off to the Hardware Store

Grades 4–5

Goals

- Determine units of measure on the basis of the measured attributes
- Recognize the need for measuring in standard units
- Become familiar with units in commonly used systems

Prior Knowledge

Students should have experience with the measurable attributes length, area, perimeter, volume, capacity, weight, temperature, and time.

Materials and Equipment

- Sales catalogs, flyers, or brochures from building supply companies or hardware stores
- One copy of the blackline master "Off to the Hardware Store" for each student

Classroom Environment

The students work in pairs, but each student completes the activity sheet "Off to the Hardware Store" independently. (Alternatively, teachers might assign the activity as a home learning project, with each student visiting a hardware store to complete the activity sheet, or as a class project involving a field trip to a local store.)

Activity

Engage

Tell your students that they are going to investigate the role that measurement plays in the buying and selling of items in hardware stores. Ask the students to create a list of the various measurement attributes that might be of interest to people who operate hardware stores or who shop at them. Students should think about length, area, volume or capacity, weight, and perhaps temperature. Ask them, "Can you think of something in a hardware store that is sold by weight?" Write their answer as the first item on a list that the class will make together. If your students are unable to make suggestions, you might propose such items as nails, washers, birdseed, or grass seed.

Focus on the attributes that hardware stores use in selling particular products. For example, they market carpet by the square foot or yard; therefore, the attribute that they use to sell carpet is area. Explain to your students that sometimes they may need to know the measure of a second attribute of an item, such as its area, perimeter, or volume, in addition to the attribute that the item is sold by, in order to buy the right amount. For example, fiberglass insulation is typically sold by length, but buyers may also need to know how many square feet a roll will cover—its area. The packaging provides this additional information.

"All students should ... understand the need for measuring with standard units and become familiar with standard units in the customary and metric systems."
(NCTM 2000, p. 170)

pp. 110–11

Explore

Arrange your students in pairs, and distribute to each pair a sales catalog, flyer, or brochure from a hardware store or building supply company. Give each student a copy of the blackline master "Off to the Hardware Store." This activity page asks them to list items that they find that are sold by length, area, perimeter, volume, capacity, weight, or any other measurable attribute. For each item, they must give the unit of measure that is used in marketing it and record the page number of the listing. Before the pairs of students begin work, show them how measurements are listed in the flyers and catalogs. The attributes may appear as headings of columns that give the units used to measure the items. For example, sterilized sand for children's sand boxes, often sold as "play sand," is typically marketed by weight and listed in catalogs in pounds:

Play Sand

Weight	Mfr. No.	Stock. No	Price Each
50 lb.	1113-51	169803	$2.86

Invite the students to begin looking through the catalogs and other sales materials to find items to list on the blackline master. Tell them that they may find measures that are not specified on the activity page, and they should list these under "Other attribute." For example, roofing shingles are sold by the *bundle* or *square*. Have the students investigate these or any other terms that they find.

Students should find three items for each attribute listed on the activity page. Be sure that they are focusing on the unit of measure that is used in marketing the item. For example, students should write the item "lumber" in the first column on their activity sheet under the attribute "Length." In the second column, "What unit of measure is used to sell the item?" they would write the particular unit of length, such as inches or feet. They should not record all the different widths and lengths of lumber that are available.

Extend

When the students have completed the activity, ask them to share with one another the items that they have listed on their activity sheets. Invite them to talk about the measurement unit and the item and to tell why they think the item is sold by that unit. Students can also explore

50 lb

Sand

$2.86

Stock Number 169803

the types of items that are sold by metric measure and the frequency with which one system of measurement is used in the hardware store as compared with another.

Assessment Ideas

You might ask your students to select one of the items that they have listed on the blackline master "Off to the Hardware Store" and tell how stores use a particular measurable attribute of the item in selling it to buyers. Ask the students what other measurements, if any, they might need to know in buying the item, depending on how they intended to use it. For example, students choosing lumber from their lists might explain that lumber is sold by length, and they could demonstrate how it is measured. They could describe a task for which a particular quantity of wood would be necessary. If they were building a bookshelf, for instance, they would know how much wood they needed to make a particular number of shelves of a particular length to fit in a particular space in a room.

List the items found for the attributes below.	What unit of measure is used to sell the items?	Page #
Length		
Sheared Fraser fir	feet	pg. 48 in flyer
Wood bases	feet	pg. 16 in flyer
Fat max tape measure	inch/feet	pg. 26 in flyer
Area		
floor tile	square inches	pg. 1245 book
Ceramic wall Tile	square inches	pg. 1247 book
Tile back board	square feet	pg. 1249 book
Perimeter		
Palm/Mist Area rug	feet	pg. 36 in flyer
Lions Heart Area Rug	feet	pg. 36 in flyer
Ivery Modern Rug	feet	pg. 36 in flyer
Volume		
microwave	cubic feet	pg. 13 flyer
Watt with Turntable	cubic feet	pg. 12 flyer
Watt Over the Range	cubic feet	pg. 13 flyer
Refrigerator	cubic feet	pg. 13 flyer
Capacity		
Fibered Roof Coating	gallons	pg. 559 book
12 year Gas water heater	gallons	pg. 1013 book
Grout and Tile Sealer	gallons	pg. 1253 book
Grout and Tile Sealer	gallons	pg. 854 book
Weight		
Bonding Mortar	pounds	pg. 1250 book
Thin Set Mortar	pounds	pg. 1250 book
White Dry Non-standing Grout	pounds	pg. 1252 book
Granite morter Mix	pounds	pg. 1251 book
Fiber glass stepladder	pounds	pg. 1 in flyer
Level Quick Underlayment	pounds	pg. 1251 flyer
fast Curing Bonding	pounds	pg. 1251 flyer

Where to Go Next in Instruction

The activity Off to the Hardware Store introduces students to standard units of measurement by showing their usefulness in daily life. In marketplace transactions, standard units of measurement are as necessary as standard units of currency, providing a common "language" that enables buyers and sellers to conduct business with one another in a manner that ensures equity, accuracy, efficiency, and satisfaction to both parties. The next activity, Measurement Scavenger Hunt, also explores the role of standard units of measurement.

Measurement Scavenger Hunt

Technically, balances are used to measure an object's *mass* rather than its *weight*. However, because the distinction between mass and weight is usually beyond the understanding of students of this age, classroom activities often use the terms interchangeably, as the activity Measurement Scavenger Hunt does here.

pp. 112–14; 115–17

Grades 3–4

Goals

- Select objects with attributes whose measurements are approximately equal to a given length, weight, or angle, as measured in standard units
- Develop a familiarity with the standard units for length, weight, and angle
- Classify measurements of objects' attributes as "the same as," "more than," or "less than" a given length, weight, or angle, as measured in standard units

Prior Knowledge

Students should have had previous experiences in using familiar devices, such as string and adding machine tape, to measure length. In addition, students should have had opportunities to use balances to measure weights and protractors to measure angles.

Materials and Equipment

- A 12-inch ruler, a 30-centimeter ruler, a yardstick, or a meterstick for each group of three or four students
- A protractor for each group of students
- A balance, with weights, for each group of students
- Assorted everyday classroom objects, such as paper clips, pencils, erasers, books, desks, and chairs distributed around the classroom
- A copy of the blackline master "Measurement Scavenger Hunt" for each student
- A transparency of the blackline master "Measurement Scavenger Hunt—Group Data" and an overhead marking pen for each group of students

Classroom Environment

The students work in groups of 3 or 4 to collect measurement data. Students complete their own copies of the blackline master, and group members work together to complete a transparency of the group's data. All students participate in the ensuing discussion.

Activity

Engage

Explain to the students that you are going to arrange them in groups to participate in a measurement scavenger hunt. Tell them that as part of their equipment for the hunt you will be giving each group a

---

This activity draws on ideas from Van de Walle (2001), pp. 291–94.

standard measurement unit for length. Without measuring, each group will then find five objects that are approximately that length. After the groups have collected their five objects, they will use their standard unit to measure and evaluate them. They will determine whether their selected objects are *the same as*, *longer than*, or *shorter than* their unit. Then they will decide which object was *closest* to their unit of length.

Show the students how the length scavenger hunt will work. Say, "Imagine that I have been assigned one foot as my unit of length for the measurement scavenger hunt." Let the students watch while you collect five objects that you think might be close to one foot in length (or you can show them five objects that you have previously collected). Then ask your students to help you measure your objects with a 12-inch ruler.

Next, distribute copies of the blackline master "Measurement Scavenger Hunt" to the students. (Show them that the scavenger hunt has three sections—on length, weight, and angle—and explain that they will be doing these one at a time. Right now, they are looking only at the section on length.)

On the board, make a chart like the first one that appears in the blackline master, shown here as figure 2.1. Use it to report on your objects' lengths. (Alternatively, you can use a transparency of the blackline master "Measurement Scavenger Hunt—Group Data" to record your results and show them on an overhead projector.) List your objects, and for each one enter a checkmark to evaluate its length in relation to a standard unit of one foot. Ask the students to help you consider each object in turn. Is it the same length as the unit? Longer than the unit? Shorter than the unit?

Object	The same length as the unit	A little longer than the unit	A little shorter than the unit
1.			
2.			
3.			
4.			
5.			

Fig. **2.1**.

Students can complete the chart to show the lengths of five classroom objects in relation to a given standard unit of length

Explain the students' goal: "After you find five objects with lengths that you think are close to one of your standard units, your task will be to find which object is closest in length to the unit that your group received."

Give a transparency of the blackline master "Measurement Scavenger Hunt—Group Data" to each group, and tell the students, "You'll make your own record of your group's results on your own activity pages, and you'll use this transparency to make a single record of your group's data that we can look at and discuss together."

Assign one of the following units to each group of students: 1 centimeter, 1 inch, 1 yard, and 1 meter. Then send your students on the length portion of the scavenger hunt.

When all the groups have completed this stage of the activity, gather the students together as a class to discuss their findings. Call on each group to put its transparency on the overhead projector, report on its measurement unit, and show its list of objects of approximately that length. When your students report on their objects, they should explain how and why they classified their lengths as the same as, longer than, or shorter than their given standard unit of length. For example, a group of students might respond, "We were measuring with a yardstick, and we decided that the top of the computer printer was about half a yard long because about as much of the yardstick stuck out beyond the printer as went across its top."

Explore

The students are now ready to continue the Measurement Scavenger Hunt, this time finding five objects with weights that are close to that of a given standard unit. Again, model the activity for your students before they begin work.

Say, "Imagine that I have been assigned one gram as my standard unit of weight for the scavenger hunt." Let your students watch again while you collect five objects that you think are all close to one gram in weight (or, as before, show your students five objects that you have previously collected). Take a paper clip as one of your objects. Using a balance, weigh the paper clip and report its weight. (A paper clip often weighs about one gram.)

Make another chart on the board, this time like the second one that appears in the blackline master, shown here as figure 2.2. (Or use a transparency of "Measurement Scavenger Hunt—Group Data" on an overhead projector.) List your objects and for each one enter a check-mark to assess its weight in relation to one gram. Ask the students to assist you in evaluating each object in turn. Is it the same weight as the unit? Heavier than the unit? Lighter than the unit? Which object is closest to one gram?

Fig. **2.2.**

Students can complete the chart to show the weights of five classroom objects in relation to a given standard unit of length

Object	The same weight as the unit	A little heavier than the unit	A little lighter than the unit
1.			
2.			
3.			
4.			
5.			

Assign one of the following units to each group of students: 1 ounce, 1 pound, 1 gram, or 1 kilogram. Then let the students complete this segment of the scavenger hunt and discuss their results.

Extend

Following the same procedure, ask the students to find five objects that have a given angle measure. (You may need to review how to measure angles with a protractor.) Again, you can model the activity by finding five objects with an angle of 90° and using a protractor to measure the angle. For example, the corner of a picture frame forms a 90° angle, as does the corner of a book or an index card.

Report your results on the board in a chart like the third one in the blackline master, shown here as figure 2.3. (Or use a transparency of "Measurement Scavenger Hunt—Group Data" on an overhead projector.) Assign one of the following angle measures to each group of students: 30°, 45°, or 60°. Send the students on the last segment of the scavenger hunt.

Object	Has an angle that is the same size as the given angle	Has an angle that is a little larger than the given angle	Has an angle that is a little smaller than the the given angle
1.			
2.			
3.			
4.			
5.			

Fig. **2.3.**

Students can complete the chart to show the measure of an angle in five objects in relation to a given standard angle measure

When students have completed all the tasks in the scavenger hunt, gather them together to discuss their findings. To guide the discussion, pose such questions as—

- "How long is a foot? A meter?" (Students should use physical referents in their replies: "A foot is about the same length as a sheet of paper"; "A meter is a little bit longer than two of our desks.")
- "How heavy is a gram? A kilogram?" (Continue to encourage students to refer to the attributes of the objects measured).
- "What does a 45° angle look like?" "A 60° angle?" (Have students use physical referents when describing the measures.)

Assessment Ideas

Principles and Standards for School Mathematics calls for students to be able to select and apply standard units in measuring attributes. Students should be given the opportunity to use a variety of tools to measure objects and to develop a familiarity with standard units. In the scavenger hunt, students measure objects that are approximately equal to their assigned unit, and they should be encouraged to use those objects subsequently as benchmarks for that unit.

For example, students might find that each side of a square floor tile in their classroom is about one foot long. Students could then use floor tiles to make approximations of lengths of other objects in the classroom. As a result of this activity, students should be ready to develop a sense of the magnitudes of several standard units.

It is important to observe how students report the measures in the scavenger hunt and how they measure the attributes of objects. Students should also be developing judgment about appropriate units for them to select in particular contexts. To help them consider this aspect of measurement, you might ask, for example, "Is a gram an appropriate unit for measuring the weight of an automobile? Is there a more appropriate metric unit? If so, what is it?"

Where to Go Next in Instruction

By grade 5, students should be ready to measure a variety of angles. To begin exploring the process of measuring angles, they can make a nonstandard protractor from wax paper or tracing paper (or another transparent material, like thin transparent plastic sheets) and use this tool in estimating angle measure.

Using benchmarks and standard units helps students extend their understanding of the measurement process. Measurement sense continues to develop throughout the later grades. Using physical referents and benchmarks and making estimates of measurements contribute to this development. It is very important to give students sufficient opportunities to refine their "unit sense."

The scavenger hunt presented in this activity allows students to develop a sense of the magnitude of selected standard units. They can now use the physical referents that they found as measurement benchmarks when they are estimating measurements in the future. The next activity, My Benchmarks, provides additional opportunities for students to explore measurement benchmarks.

My Benchmarks

Grades 4–5

Goals

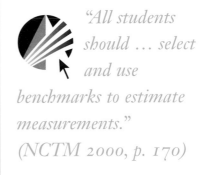

"*All students should … select and use benchmarks to estimate measurements.*"
(*NCTM 2000, p. 170*)

- Generate ways of using familiar objects as measurement devices, or *benchmarks*, in daily life
- Use these benchmarks to estimate the measurements of selected attributes of familiar objects
- Develop a familiarity with standard units by measuring particular attributes in familiar objects

Prior Knowledge

In grade 3, students should have had experience in using familiar devices, such as string, adding machine tape, or other tools, to measure the distance between two points. In addition, students should be able to use scales to measure weight and measuring cups to measure volume.

Materials and Equipment

- A classroom set of measurement instruments, such as 12-inch and 30-centimeter rulers, scales, measuring cups
- Five or six rolls of adding machine tape
- A one-meter length of rope
- A dozen empty containers (one-gallon milk jugs and one-pint bottles)
- Three or four one-pound bags of flour or sugar, a one-gallon jug of water, and other items that are measured by volume, weight, or capacity
- A list of ten to fifteen items that students can measure in the classroom, school building, or playground, such as chalkboards, books, and door frames
- A copy of the blackline master "My Benchmarks" for each student
- A sheet of newsprint for reporting students' results

pp. 118–21

Classroom Environment

The students work in pairs but complete activity pages independently.

Activity

Engage

Explain to your students that they will be generating a set of *benchmarks*. Benchmarks are personal referents that one can use to estimate measurements such as length, area, volume, or weight, in standard units.

Ask your students to select five items from your list of objects that they can measure in the classroom, the school building, or the playground.

This activity draws on ideas from Van de Walle (2001), pp. 292–93.

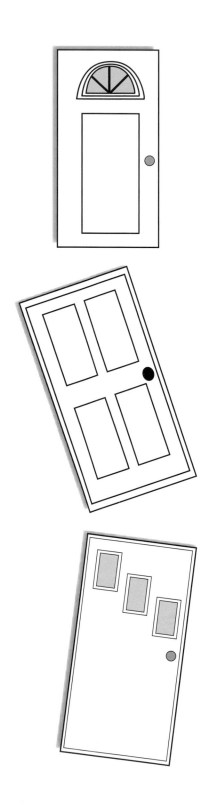

Tell them to list their items and choose a measurable attribute for each one.

Distribute copies of the blackline master "My Benchmarks." The worksheet provides space for you and your students to work an example together. In the example, a doorway is the object to be measured. Guide the students through the task with this object. Here the attribute to be measured is height. Ask the students about appropriate units, both customary and metric, for measuring the height of the doorway to their classroom. Pose such questions as—

- "Would millimeters be appropriate for measuring the doorway?"
- "Would centimeters be appropriate?"
- "Would we want to make this measurement in inches?"
- "Would we want to use feet?"

Always ask your students to explain their responses.

Next, turn the discussion to benchmarks. Discuss possible benchmarks that the students might use to arrive at an estimate, in standard units, of the height of the doorway. You might suggest a one-meter length of rope or a one-foot sheet of construction paper. Have these or similar benchmarks ready to show your students, and use them together in estimating the height of the doorway.

Emphasize that the students should use the language of approximation when they are reporting estimates: "The door is *about* 2 meters high"; "Its height is *around* 7 feet." When students make an estimate, ask them to explain how they arrived at that number. For example, a student might say, "It looked to me as though I could fit two lengths of rope along the side of the doorway, so I think the height is around *2* meters."

Next, use a yardstick and a meterstick to make actual measurements of the height of the doorway in both customary and metric units. You can make the measurements yourself, with the students looking on, or you can ask a pair of students to make them while their classmates evaluate their measuring techniques.

Repeat this process of estimation, this time using some other material listed for the activity, such as a one-pound bag of flour or a one-gallon jug of water. For example, you might ask students to help you use a bag of flour to estimate the weight of a stack of books. After they have made their estimate, you should weigh the books to see how close the estimate was to the actual weight of the stack.

Explore

Arrange the students in pairs to perform the tasks on the blackline master "My Benchmarks." Although students work in pairs, each student should complete his or her own activity pages. The students will be using benchmarks to estimate a measurement for each of the five objects that they selected earlier from your list of suitable objects in the classroom or school building or on the playground.

Note that the students will be using customary units for some of their estimates and metric units for others. They will be checking all estimates against actual measures of the items' attributes.

When all the pairs of students have completed the activity, ask each pair to share the benchmarks that they used to estimate the measures.

On the board or a sheet of newsprint, record the benchmarks that the students used to estimate various measures in standard units. Be sure to have the students report how they used the benchmarks to arrive at their estimates.

Extend

As a home learning activity, students can use their benchmarks to estimate measurements for five objects in both metric and customary units. They should measure a different attribute—length, weight, area, volume, or capacity—in each object. After they have made their estimates, students should use conventional measurement instruments to check them. Ask students to organize their findings in charts like those shown in figure 2.4.

Using Benchmarks to Estimate Attributes in Five Objects— Customary Measurements				
Object	Attribute	Benchmark	Estimate	Measure-ment
1.				
2.				
3.				
4.				
5.				

Using Benchmarks to Estimate Attributes in the Same Five Objects— Metric Measurements				
Object	Attribute	Benchmark	Estimate	Measure-ment
1.				
2.				
3.				
4.				
5.				

Fig. **2.4.**

Students can complete charts like those shown to organize their findings

You might also ask your students if they could use any of the objects that they just measured as benchmarks in the future. For example, if they knew that the classroom doorway was two meters high, could they use it as a benchmark to estimate the height of the classroom ceiling?

Assessment Ideas

Principles and Standards for School Mathematics calls for students to be able to select and use benchmarks to estimate measurements. Through the repeated use of benchmarks, students develop a feel for customary and metric units. In addition, experiences with estimation help students internalize the measurement process. As a result of activities like the previous one, students should be able to use attributes of familiar objects as benchmarks to estimate measurements of other objects in standard units.

Throughout the discussion, observe how your students report their findings. Ask if other pairs of students used the same type of object as a benchmark for a particular measurement. If so, do they agree or disagree that the benchmark established by the first pair is appropriate? Encourage students to estimate measures again if there is disagreement. This exercise provides an opportunity for students to engage in both peer- and self-assessment and allows for additional development of measurement sense.

Where to Go Next in Instruction?

Children in grades 3–5 need many additional experiences to reinforce and build on their first experiences in using benchmarks to estimate measurements. One question to examine in these subsequent investigations is "What degree of precision should a given measurement have?" The next activity, How Precise Should My Measurement Be? gives students an opportunity to explore this question.

How Precise Should My Measurement Be?

Grades 3–5

Goals

- Identify measurements in everyday activities
- Understand that measurements are approximations and that differences in units affect precision

Previous Knowledge

In prekindergarten through grade 2, students should have had opportunities to measure the attributes length, area, volume, capacity, or weight in a variety of objects and to record their measurements.

Materials and Equipment

- A copy of the blackline master "My Measurement Activities" for each student
- Students' mathematics journals

Classroom Environment

Students make entries in their mathematics journals at home each day for a week and discuss them in class at the end of that time.

Activity

Engage

Invite your students to consider two hypothetical situations involving measurement:

1. "Suppose that you are helping your friend measure a window frame for a new pane of glass. Working together, you and your friend measure the length and width of the frame so that you can order the glass."

2. "Suppose that you have learned that it is harmful to carry a very heavy book bag. You have decided to put a limit of ten pounds on the weight of the books and other materials that you will carry in your bag."

Ask the students, "Which measurement should be more precise—the dimensions of the window frame or the weight of the books? Why?"

Students need to understand the difference between how precise a measurement *needs to be* and how precise it *can be*. The *context* and *purpose* of a measurement are factors that determine how precise a measurement *needs to be*. It is important that the dimensions of the window frame be measured precisely so that the pane of glass will fit snugly. It is less important that the weight of the books be a precise amount. It can be a little under or a little over the 10-pound mark without causing any harm.

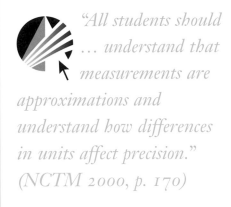

"*All students should … understand that measurements are approximations and understand how differences in units affect precision.*" (NCTM 2000, p. 170)

pp. 122–23

The degree of precision that a measurement requires depends on the context and purpose of the measurement.

Explain to your students that a factor in how precise a measurement *can be* is the *size of the unit*. A small unit generally produces greater precision than a large unit. The dimensions of the window frame may be given in feet and inches (smaller unit), down to small fractions of an inch, so that the person who is going to cut the glass can work with a very precise measure. By contrast, the weight of the books may be measured to the nearest whole pound, since precision in this context is less important.

Explore

Tell your students that they will explore these ideas about how precise measurements need to be (depending on the context and purpose) or can be (depending on the unit size) by compiling a list in their mathematics journals of the activities involving measurement that they do during a week.

Distribute copies of the blackline master for "My Measurement Activities" (see the journal portion in fig. 2.5). You can have your students use it as a model for setting up pages in their journals, or they can paste it into their journals. Another strategy would be to have your students complete the activity page over the course of a week as a freestanding measurement journal, separate from any other mathematics journal.

To stimulate thinking for the journal entries, ask the students to brainstorm about activities that they do inside and outside school during a typical week. Ask them to think about the following questions as they consider their activities:

- "What measurements do you make before school?"
- "What measurements do you make during the school day?"
- "What measurements do you make during after-school activities or at lunchtime?"
- "What measurements do you make when you go shopping or plan a shopping trip?"
- "What measurements do you or other members of your family make in preparing meals?"

At the end of a week, when your students have completed the lists in their journals, ask them to report their measurement activities. For each activity that they identify, ask them to talk about both the measurement unit and the tool that were appropriate for the measurement. For example, a student might report that he uses two cups of water to make his noodle soup for his after-school snack and that if he doesn't use exactly two cups his soup "doesn't turn out right."

Ask the students to look at their lists and talk about which measurements had to be precise and which ones did not. Ask them to think about, discuss, and write the reasons for their choices.

Students may be forming the idea that many of the measurements that they make are approximations. The measurements do not "come out evenly." Encourage students to search for smaller units with which to refine their measures. At this point, it is important to discuss with students the fact that measurements of continuous quantities (such as length) are always approximations and, as such, include some error. Although measuring with smaller subdivisions of units allows for greater precision in measurement, even the smallest units and most

Fig. **2.5.**

A portion of a sample journal for students to complete, as shown on the blackline master "My Measurement Activities" precise measurements that we can devise will include some error, since there is no *smallest* measurement unit.

Extend

Ask the students to use rulers that have millimeter markings to measure the lengths of the edges of selected pattern blocks. Tell the students to report their measurements to the nearest millimeter. Have the students post their results on the board or a transparency and then discuss any differences in the measurements.

When you discuss the differences in the students' results, introduce the idea of *accuracy* (and *inaccuracy*) in measurement. Point out that differences in measurements can result from variables in the situations in which the measurements are made. These variables include the different people who make the measurements, the different conditions under which they make them (temperature, lighting, etc), the different instruments that they use, and the different techniques and procedures that they follow.

Explain to the students that certain ways of making measurements have become standard and that these accepted ways ensure that results will be as uniform as possible. We have to learn these ways of measuring and make an effort to follow them. For example, we always try to read measurements made with a ruler "straight on" instead of from what we think of as a "bad" (distorting) angle of vision.

Emphasize that the differences that we often see in multiple measurements of the same object reflect errors that are different from the unavoidable or inherent error that you were discussing earlier. Usually, differences in multiple measurements are avoidable. They do not reflect the inevitable imprecision of all measurements.

Tell your students that when they see differences in their classmates' measurements of the same object, they are seeing examples of what we usually talk about as the accuracy or inaccuracy of measurements. We recognize that a measurement of a continuous quantity can be inaccurate as a result of errors in the way in which it was made at the same time that it is always and necessarily imprecise, as a result of the error that is inherent in measurements made with any instrument.

Make sure that the students understand the important idea that real-world measurements of continuous quantities are always approximations. It is not possible to guarantee that any object is exactly one meter long; most centimeter rulers can measure only to one-tenth of a centimeter (a millimeter), for example.

Assessment Ideas

Over time, students should be given opportunities to reflect on the measurement activities that they do and to discuss the appropriateness of using the language of approximation when stating measurements. Students should also have numerous experiences in using subdivided units to introduce greater degrees of precision into their measurements. Observe how the students discuss their lists of their measurement activities, the appropriateness of their measurement processes, and the appropriateness of the relationship that they make between the precision with which they are measuring and the context of the measurement. Many students will need repeated exposure to these ideas before they understand them fully.

Pattern Blocks 3

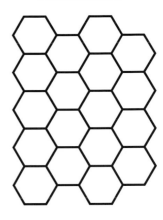

Pattern Blocks 5

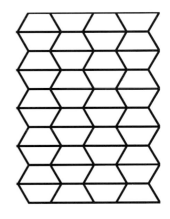

 Templates for pattern blocks appear on the accompanying CD-ROM.

Where to Go Next in Instruction?

Students should become proficient in using conventional measuring tools to produce accurate measurements in a variety of contexts. In the process, students may encounter the need to convert from one standard unit to another within a measurement system. For example, students may determine the weight of an object to be 75 ounces and may need to know how many pounds are in that many ounces. They may know that one pound equals 16 ounces but still not be sure how to determine how many pounds are in 75 ounces. This chapter's final activity, Conversion Sense, introduces unit conversion in the context of measuring time.

Conversion Sense

Grades 4–5

Goals

- Carry out simple conversions from one standard unit of time to another
- Solve problems applying unit equivalences for measures of time

Prior Knowledge

Students should be able to tell time on an analog clock or watch and measure time with a second hand. Students should also be familiar with time equivalences, such as that 60 minutes equal 1 hour.

Materials and Equipment

- A copy of the blackline master "Conversion Sense" for each student
- 100 pieces of paper, cut uniformly with a paper cutter to the size of dollar bills
- One calculator for each pair of students
- One stop watch

Classroom Environment

The students work in pairs on the problems on the blackline master and come together as a class to discuss the solutions.

Activity

Engage

Pose the following problem to students: "How much time would it take to count one million one-dollar bills?" Ask them how they would go about solving this problem. Some students might suggest that it would take about one second to count each dollar bill. Explain to the students that in order to answer this question, they would need to apply their knowledge of time equivalences.

Discuss possible strategies for solving the problem. For example, one student might say, "We have to find out how long it takes to count one hundred one-dollar bills." Another might offer, "Once we have this information, we will need to divide one million by one hundred to find out how many sets of one hundred are in one million."

To determine how long it would take to count 100 one-dollar bills, select two volunteers, each of whom will perform a trial. Use a stopwatch to time these students as they count one hundred dollar-bill-sized pieces of paper. (Cut these pieces in advance on a paper cutter, making them as uniform as possible.) Have the two students perform their trials one after the other. (You may want to send the one who is going to count second out of the room while the first student counts so

---

This activity draws on ideas from Reys et al. (2004), pp. 410.

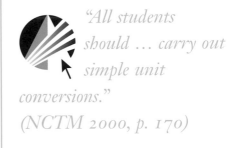

"All students should … carry out simple unit conversions."
(NCTM 2000, p. 170)

p. 124

that the second student won't make a special effort to "beat" the first student's time!) If the times are different for the two students (and they may be), average the two. Explain that the average gives an estimate of how much time it would take to count one hundred one-dollar bills.

Explore

Distribute the blackline master "Conversion Sense" to the students. Arrange them in pairs, and give them time to work on the problem with their partners, using calculators if they wish. Then convene the students as a class to discuss their solutions. One solution follows (your students may come up with other times):

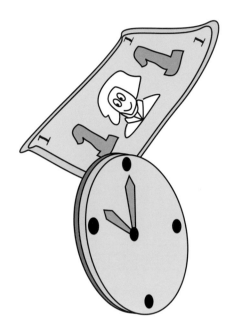

"Our trials show that 100 one-dollar bills can be counted in 60 seconds. Sixty seconds equal one minute. Since there are 60 minutes in one hour, we can multiply 100 by 60 to find out that we could count 6000 one-dollar bills in one hour (assuming that we didn't slow down over time).

"We know that there are 24 hours in one day, so we can multiply 6000 by 24 to find out that we could count 144,000 bills in one day (assuming that we didn't slow down or sleep). Now we can divide 1,000,000 bills by 144,000 bills per day to find out how many days it would take to count one million one-dollar bills. We get approximately 6.9 days. Therefore, it would take someone who was counting around the clock about 7 days to count one million one-dollar bills at a constant rate of 100 bills per minute."

After the class has completed its discussion of this problem, ask the students to work in pairs again, this time to do the heartbeat exploration that appears on the blackline master as problem 2: "How many times does your heart beat in one-half of an hour?" The students work with their partners first to develop a strategy for answering the question and then to compute the answer, which they should be ready to explain in detail.

Paired students might decide to count each other's pulse for 10 seconds. (They might ask to borrow the stopwatch to do this if the classroom doesn't have a clock with a second hand. Or one of the students might have a watch that shows seconds.) Let's say that a pair of students averaged their counts and arrived at 12 beats in 10 seconds. Then they might reason as follows:

> "We need to figure out how many heartbeats that would be in a minute. Okay, one minute equals 60 seconds, and 60 seconds equals 6 times 10 seconds. So one minute has six 10-second stretches of time. We know that each 10-second stretch of time is going to have 12 heartbeats if the pulse keeps going at the same rate. So in one minute there will be 12 times 6, or 72, heartbeats. Okay. Now, one-half of an hour equals 30 minutes. So in half an hour there would be 30 times 72, or 2160, heartbeats."

Extend

Unit conversions are most meaningful when students make them in the context of problem-solving situations such as those in these two problems. These contexts have the additional advantage of presenting openings for connections to other, related topics, such as problem-solving strategies, data analysis, and pattern recognition. To give the students additional practice, create problems involving the conversion of units of—

- weight ("How much would 1 million dollars weigh?");
- volume ("How many liters of coke could I pour into an empty aquarium like the one in our classroom?"); and
- area ("How many baseball cards could cover our classroom floor?").

Assessment Ideas

Principles and Standards for School Mathematics calls for students to be able to carry out simple unit conversions. Observe students as they solve the problems on the blackline master "Conversion Sense." Some students may already be familiar with many unit equivalences and be comfortable in using them. Students who have not previously worked with equivalences may need further instruction or experiences with unit conversions.

Where to Go Next in Instruction?

Students' understanding of unit conversions will give them a background and conceptual support for working with rates. The second question on the blackline master, "How many times does your heart beat in one-half of an hour?" is an example of a problem in which an understanding of time units and their equivalences facilitates a solution involving a rate.

This chapter has provided a transition from measuring attributes of objects with nonstandard units to measuring them with standard units. The activities have focused on the idea of benchmarks to facilitate the estimation of measurements as well as the notions that all measurements of continuous quantities are approximate and that context, purpose, and unit size play important roles in determining how precise a measurement has to be or can be. The following chapter transfers these fundamental ideas to the context of two-dimensional shapes in an exploration of perimeter and area.

NAVIGATING *through* MEASUREMENT

Chapter 3
Measuring Two-Dimensional Shapes

Exploring two-dimensional shapes gives students essential opportunities to deepen and expand their measurement ideas and skills as *Principles and Standards for School Mathematics* (NCTM 2000) recommends. The activities in this chapter allow students to investigate and understand the attributes *area* and *perimeter*, which are important focuses of measurement instruction in grades 3–5. Activities that enable students to make sense of their experiences with area and perimeter should help them achieve three broad goals:

1. Develop an understanding of the meaning of area and perimeter;
2. Be able to recognize real-life applications of area and perimeter; and
3. Discover the formulas for calculating area and perimeter efficiently.

The chapter begins with Big Cover-Up, an activity in which students make sense of the attribute *area*. They figure out effective ways of estimating, counting, and comparing the areas of two or more objects, and they explore appropriate units for measuring the area of a particular object. Once students have explored the areas of many objects through informal methods, such as covering and counting, they will be ready for a more formal approach to finding the area of rectangles.

Principles and Standards also recommends that students apply "appropriate techniques and formulas to determine measurements" (p. 170).

The second activity, Stuck on Stickers, asks students to imagine that they are the owners of a sticker factory, and they must determine the dimensions of rectangular sheets of stickers that their factory will sell. They need to find a formula for the area of the sheets to help them with their work. Through this role playing, students make sense of the *length × height* formula for the area of rectangles and link this new information to their previous learning about multiplication arrays.

In the third activity, Changing Garden, students explore the attribute perimeter and discover the formula for the perimeter of a rectangle as they work with a realistic scenario. Students determine all possible ways to arrange 30 feet of fencing around a garden with whole-number dimensions. Students also explore how changing the dimensions of the garden affects its area. Through their work, the students discover the links between the dimensions of a rectangle and its area.

Finally, the two activities Geo-Exploration—Parallelograms and Geo-Exploration—Triangles use geoboards to engage students in discovering formulas for calculating the area of parallelograms and triangles. Students make sense of the *base × height* formula for the area of a parallelogram by exploring the similarities between parallelograms and rectangles. Then they develop the formula 1/2(*base × height*) for the area of a triangle by linking triangles to parallelograms. By making connections among the shapes, students make connections among the formulas for their areas. Throughout the explorations, students see the formulas as grounded in mathematics and not as arbitrary rules. They develop an understanding of the formulas as efficient and meaningful ways to find and compare areas of two-dimensional shapes.

Big Cover-Up

Grades 3–4

Goals

- Describe the area of any noncurved two-dimensional shape (regular or irregular) as the number of square units needed to "cover" the shape
- Choose an appropriate unit to measure the area of a particular two-dimensional shape

Prior Knowledge

Students should have had previous experiences in investigating linear measurement with nonstandard as well as standard units and should be familiar with customary and metric units of linear measure.

Materials and Equipment

- A copy of the blackline master "Big Cover-Up" for each student
- A copy of the blackline master "Big Cover-Up Goes Home" for each student
- Five or six sheets of cardstock or construction paper
- Five or six pairs of scissors
- A sheet of chart paper

For each pair of students, two dozen—

- Squares, triangles, and (blue) rhombi, all from pattern-block sets
- 3-by-5-inch index cards

Classroom Environment

The students work first as a whole class and then in pairs. Finally, all the students share the results of the exploration.

Activity

Engage

On cardstock or construction paper, draw two irregular shapes that are different in area. Concave polygons such as those in figure 3.1 are possible examples. Show your figures to the class, and ask, "How can we figure out which of these two shapes is larger?" Post students' suggested methods on the board and try each one. Students might suggest covering both shapes with square tiles, counting the number of square tiles, and comparing the two quantities. Another strategy might consist of cutting one shape into pieces and arranging them on top of the other shape. The students could then compare the areas of the shapes. Some students might not understand that if a two-dimensional shape is cut

---

The "Engage" section of this activity is loosely based on ideas in Cathcart et al. (2001) pp. 336.

"All students should … understand such attributes as … area … and select the appropriate type of unit for measuring each attribute; [and] … develop strategies for estimating the … areas … of irregular shapes." (NCTM 2000, p. 170)

pp. 125, 126

See the accompanying CD-ROM for a template for pattern blocks.

For common examples of students' ideas and misconceptions about area, see "Developing Spatial Sense through Area Measurement" (Nitabach and Lehrer 1996) on the CD-ROM.

and its pieces are rearranged, its area is conserved. Use scissors to cut one of your shapes to demonstrate this point.

Fig. **3.1.**

Exploring irregular shapes

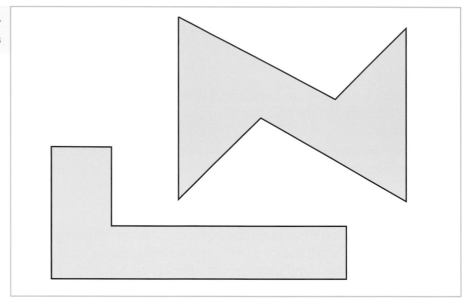

Explain that as the students have just suggested, one two-dimensional shape is considered "larger" than another if its area is greater. Mention to the students that mathematicians use the word *area* to describe the space that a two-dimensional shape occupies (also referred to as its *surface*). Note that in some instances, *size* is a useful substitute for *area*, although the terms do not mean exactly the same thing.

Explore

Say to your class, "Let's list some two-dimensional, noncurved objects in the classroom whose areas you think you could find." Possible items include a sheet of paper, a bulletin board, a poster, and a computer screen. Be sure to clarify your students' suggestions when necessary. For example, if students suggest a desk, remind them that area applies only to two-dimensional shapes. The desk is three-dimensional but has surfaces, and these are two-dimensional, with areas that can be measured.

List all the objects on a sheet of chart paper. Distribute the blackline master "Big Cover-Up." Explain to the students that they will work in pairs to select two objects on the list and measure their areas. Later they will share their findings with the whole class.

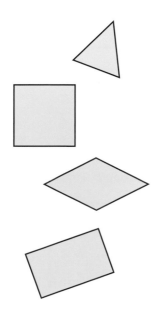

Tell the students that they will measure each object with several different nonstandard units. Review the term *unit* if necessary. Explain that for this activity, the units will include square tiles, triangles, and rhombi from pattern-block sets, and 3-by-5-inch index cards. Students should complete the chart on the blackline master as they work.

Choose one item from the list, and model the process for the class. Figure 3.2 shows a partially completed chart like the chart on the blackline master. Set up such a chart on the board or an overhead transparency. Discuss why some measurements on the chart will be fairly exact and others will be clearly estimates. (Though all measurements are approximations, they will be more exact if a whole number of units fits on the object without gaps or overlaps. Measurements will be less exact and useful only as estimates if the units that cover an area leave gaps that only fractional pieces of the unit can fill.)

Object	Number of Triangles	Number of Rhombi	Number of Squares	Number of Index Cards
1. Cover of an assignment notepad	A little more than 50	A little more than 25	24 exactly	One and a little bit more (but closer to one than two)

Let the students watch while you cover the object that you have selected, first with triangles, then with rhombi, squares, and 3-by-5-inch cards (rectangles). Be sure that you have chosen an object that will allow you to show the students what to do if they use all the units of one type in their own collection before they have finished covering the object. Suggest that they try to borrow units from another pair of students to complete the task. Or they can use a pencil and mark lightly on their shape to show the boundary between the area that they have already covered with units and the area that they still need to cover. Then they can move their units and reuse them to cover a new area of their shape.

As you measure the area of your item with units of each type, record your finding in your chart. When you are finished, leave your chart for the students to refer to as they record their findings, and let them begin the activity.

When the students have completed the blackline master, call them together as a class, and give them an opportunity to share their strategies for finding the areas of their shapes with the different units. Facilitate their sharing by asking questions such as the following:

- "Did anybody use estimation to figure out how many units would fill the gaps that were too small to cover with a unit?" (Discuss with the students how they could have combined small gaps until their areas equal the area of one of the units being used to measure the shape.)

- "Did anybody use symmetry to find the area of an object?" (Review the terms *symmetry*, *symmetrical*, and *line of symmetry* with the students. Discuss the fact that some two-dimensional shapes have symmetry and others do not. Ask the students if either of their shapes was symmetrical. If so, ask, "How could you have used this fact to find its area more quickly?" (They could have divided the area along the line [or lines] of symmetry, covered a section with their units, and multiplied by the number of symmetrical sections.)

- "Did anybody use multiplication in another way?" (If the students could divide a shape into sections that were not symmetrical but were still equal in area, they could have figured out how many

See Navigating
through Geometry
in Grades 3–5
(*Gavin et al.* 2001,
pp. 47–74) *for a
discussion of concepts
of symmetry that are
important for students
to understand in the
upper elementary
grades.*

A closed figure like the one
shown here has
no symmetry,
but students can
divide it into
equal sections,
cover one with
units of area,
count the
units, and
multiply the
number by
the number
of sections
to find the
area.

units they needed to cover a section of the shape and then multi-plied to figure out how many of those units they would need to cover the whole shape.)

- "Did anybody notice a relationship between measures of area made with triangles and those made with rhombi?" (In sets of pattern blocks, the area of a triangle is half the area of a blue rhombus. So the number of rhombi that students need to cover an area should be half the number of triangles. Or, conversely, the number of triangles that they need to cover an area should be twice the number of rhombi.)

- "Of all the objects that you measured, which had the greatest area? Which had the smallest area? How can you know?" (The shape with the greatest area is the one that required the greatest number of units of a particular type to cover it. The shape with the smallest area is the one that required the smallest number of the same units to cover it. In other words, to answer the question, the students should look at all the measurements that they made in any one type of unit—triangles, for example. The largest object will be the one that required the greatest number of triangles to cover it, and the smallest will be the one that required the smallest number of triangles to cover it.)

To help students reflect on the measurements that they have made, turn the discussion to questions 2 and 3 on the blackline master:

- "What relationship did you notice between the size of each unit and the total number of units that you needed to cover the shape?" (Students should have noticed that the larger the unit, the smaller the number of units needed to cover the object. This inverse relationship between the size of a unit and the number of units is worthy of continued discussion and emphasis throughout the measurement unit.)

- "Are your measurements fairly exact or are they useful mainly as estimates? How do you know?" (Many of the students' measure-ments will not be extremely precise since their units and tech-niques will often leave gaps whose areas they must either omit or estimate.)

Extend

Help the students make a transition from nonstandard to standard units by asking, "What are some other units that we use to measure area?" (Students might suggest a square inch, square foot, square yard, square mile, square centimeter, square meter, or square kilometer.) Emphasize the importance of thinking in terms of two dimensions. Provide several pairs of scissors and sheets of construction paper or cardstock, and let your students collaborate as a class to create and post models of selected square units.

Let the class explore a real-life application of the covering method of measuring area. For example, help students cut paper to a square foot to measure the area of a bulletin board. Ask your class to brainstorm to make a list of tasks that require finding the area of a shape. Possible tasks include determining the area of —

- a floor, to install wall-to-wall carpet;
- a backyard, to decide whether or not to buy a house;
- a park, to lay out a baseball diamond; and
- a wall, to hang wallpaper.

For each job, ask, "Can we use an estimate, or do we need a precise measure?" Be sure that students remember that a measurement's context and purpose—that is, the situation in which the measurement is made and the use that someone plans to make of it—determine how precise the measurement needs to be. For example, if people wanted to order wall-to-wall carpet for a room, they would need to have a more precise measurement than if they were shopping for an area rug for the room and were comparing costs of different rugs.

After the class has listed all the tasks that students can think of that require measuring an area, ask, "Which square unit would you use to find the area in each situation? Why?" The students should see that a square foot would be appropriate for measuring the area of a floor or wall, for example, but something larger, such as a square yard or square meter, would be more appropriate for a backyard or park.

Informally assess your students' responses to see if they understand three main ideas:

1. The size of the object that they intend to measure and the size of the unit that they plan to use to measure it should be in an appropriate relationship to each other (that is, the larger the object, the larger the unit should be).
2. How precise a measurement needs to be is determined by the purpose and context of the measurement.
3. How accurate a measurement is depends on the physical process that is followed in measuring the attribute in an object.

Assessment Ideas

Distribute copies of the blackline master "Big Cover-Up Goes Home." Ask students to work at home to measure the areas of three additional two-dimensional shapes that they find there. Tell the students to use an appropriate standard square unit for each object. In class the next day, have them report on each object that they measured, giving its area, the size of the unit that they used to measure it, and their reason for selecting a unit of that size.

Where to Go Next in Instruction

Students can benefit from opportunities to estimate and measure large areas, such as the floor of the classroom, the cafeteria, or the gymnasium. Students should use models that they construct of square feet, square yards, square meters, and so on, to select units that are suitable for the spaces. Students should estimate the area of each surface before developing a plan for measuring the area. After carrying out their plan, they should share their results with one another.

As you help your students explore these area measurements, guide them to think about ways to determine area without having to cover a surface. The next activity, Stuck on Stickers, takes students toward this goal, as they develop a formula for calculating the area of rectangular shapes.

pp. 127–29

Stuck on Stickers

Grades 3–4

Goals

- Find the area of a rectangle by multiplying its length by its width
- List the length and width (in whole number values) of each possible rectangle with a given area

Prior Knowledge

Students should have had experience in comparing the areas of shapes, and they should know that the area of a two-dimensional shape equals the number of square units that fit inside or cover it.

Materials and Equipment

- A copy of the blackline master "Stuck on Stickers" for each student and a transparency of it for the teacher
- Two or three examples of commercially produced sheets of stickers
- Three dozen square-inch tiles for each pair of students
- Three or four sheets of one-inch grid paper for each pair of students
- One ruler for each student

Classroom Environment

The students work in pairs on the activity and then come together as a class to discuss their work.

Activity

Engage

Show the class several examples of still-packaged, commercially produced stickers. Point out that some stores offer stickers on sheets, some display them on rolls and allow customers to buy strips in lengths that they want, and some offer groups of stickers in envelopes. Let the students spend a few minutes sharing observations about stickers and how they are customarily packaged for sale. Sustain the discussion by asking such questions as "What shapes are these stickers?" and "How are they arranged and presented for sale?"

Sometimes stickers are arranged in rows and columns on rectangular sheets. The stickers that your students will be thinking about in this activity are of this type.

Tell your students to imagine that they are the owners of the Stuck-on-Stickers Sticker Factory, and they are in charge of designing new sheets of stickers to sell to stores. The stores that want to carry stickers from the Stuck-on-Stickers Sticker Factory have certain requirements:

1. All the sticker sheets must be rectangular, with dimensions that are whole numbers of inches—no fractions.

2. The stickers must be arranged on the sheets in rows and columns.

3. Each sticker must occupy exactly one square inch of space.

Explore

Distribute copies of the blackline master "Stuck on Stickers" to each student, arrange the students in pairs, and give the pairs square tiles and one-inch grid paper. In the first task, each pair of students, acting as co-owners of the sticker factory, will examine an order from a store that has specified dimensions for sticker sheets that it wants to sell. The students must determine how many square-inch stickers would cover each sheet.

In the second task, the students examine an order from a store that instead has specified the total number of stickers that it wants on each sheet. Here the students must find the length and the width (whole-number values only) of the sheet—or possibly sheets, since now there may be more than one answer!

After the students have completed these tasks, the activity page asks them to describe their work and then identify the biggest sheet and the smallest sheet that the factory would need to have in stock to fill the orders. Finally, the students suggest five sticker sheets (identified by their dimensions and the number of stickers per sheet) that they would direct the Stuck-on-Stickers Sticker Factory to manufacture.

When everyone has finished the activity pages, bring the class together. Use the chart on a transparency of the "Stuck on Stickers" blackline master or draw a replica of the chart on the board and ask the students to share their findings. Encourage the discussion by asking the following questions:

- "What is the difference between order 1 and order 2?" (The difference is in what information each order gives or does not give. Order 1 gives the dimensions of each sheet but not the total number of stickers that it will hold. Order 2 gives the total number of stickers that a sheet will hold but not its dimensions.)

- "How do you know that you are working with area in these tasks? (Stickers cover each sheet. Area is the attribute of an object that can be measured by covering.)

- "What is the relationship between the total number of stickers on a sheet and the area of the sheet?" (All the stickers are the same size, and they completely cover the sheet, so the area of the sheet equals the number of stickers.)

- "What could you do if you didn't have enough square tiles to cover one entire sheet? Could you find the total number of stickers by covering *part* of the sheet with square tiles? If so, how?" (One strategy would be to cover one row or column with tiles and count how many it takes to fill it. The next step would be to use tiles to count how many rows or columns would fill the entire sheet. The final step would be to multiply to get the total number of tiles [stickers] that would fill the sheet. See figure 3.3.)

- "What patterns do you see so far in your work on the blackline master? What relationship do you see between the length and width of a sheet and its area?" (The area of the sheet can be found by multiplying the length by the width.)

- "Think back to what you know about multiplication. What is the connection between our work today and what you know about

Making connections between mathematics topics leads to a greater understanding of ideas.

Fig. **3.3.**

A strategy for measuring area

Count the rows = 4
Count the columns = 3
4 × 3 = 12 stickers

See the Stuck on Stickers applet on the accompanying CD-ROM to give students additional practice with the idea that area = length × width.

multiplication arrays?" (It makes sense that the area of a sheet can be found by multiplying length by width, since a sticker sheet looks like an array of equal rows and equal columns. To find the total in an array with equal rows and equal columns, we can simply multiply the number of rows by the number of columns.)

- "Does this help you figure out a formula or method for finding the area of *any* rectangle? If so, what is the formula, and why does it work?" (The formula for the area of a rectangle is *length × width = area*. It works because a rectangle can be thought of as an array of equal rows and equal columns. The length is equal to the number of columns. The width is equal to the number of rows in each column. Multiplying the rows by the columns gives us the whole array.)

The sticker sheet shown in figure 3.3 has 4 rows. There are 3 stickers in a row, or, in other words, there are 3 columns. Since we have an array of equal rows and equal columns, we can find the total number of stickers, or the area of the sheet, by multiplying 4 times 3. Thus, the total area is 12 stickers.

Extend

The applet Stuck on Stickers on the CD-ROM lets students practice using the relationship that they have just discovered between the length and width of a rectangle and its area. Continuing the sticker-sheet investigation, the applet reproduces the main features of activity tasks 1 and 2. It can show the lengths and widths of a variety of rectangular sticker sheets and can let students find the total number of square-inch stickers that will fit on them. Or the applet can show the total number of stickers (area) on a rectangular sheet and can let the students determine the dimensions of the sheets. (Now there may be more than one sheet!)

Students can also extend their learning by exploring how to use rulers to measure the area of rectangles. Be sure that all the students have rulers, and ask them to create rectangles of various sizes in whole numbers of centimeters or inches. Tell them to find the area of each rectangle that they create. Discuss how using a ruler is similar to and different from using square tiles to find the dimensions and area of a rectangle.

Assessment Ideas

Ask students to write a letter telling a friend how to find the area of any rectangle or rectangular surface. In their letters, they should give a step-by-step explanation of how to use the formula for the area of a rectangle. They should also explain why the formula works, giving examples.

Where to Go Next in Instruction

Have your students brainstorm to create a list of real-world situations in which finding the area of rectangular figures would be useful. Discuss appropriate units of measure for each, and include these on the list. Post the students' work on a bulletin board. Examples might include determining the area of a room to purchase flooring or carpeting (square yards or square meters) or finding the area of a wall to order wallpaper (square yards or square meters). Square kilometers are appropriate for measuring large areas of land.

Throughout the year, share with your students any situations in which you need to find areas of rectangular objects, either outside school or in preparing a lesson. You can relate personal experiences, such as the following:

> "I was responsible for purchasing plastic covering for the tables at our annual family reunion. We had 14 rectangular tables of different sizes. I did not want to measure each one and add the areas, so I measured the largest table, computed its area in square feet, multiplied the result by 14, and purchased that much material. We used the leftover plastic to make banners with numbers for the tables."

Be sure to tell the students the unit of measure and the measurement strategy that you used in each situation. Encourage them to do the same.

Ask your students to choose situations from their posted list and use them in writing real-life story problems involving area. For instance, a student might write, "My mom wants to put a tool shed on a rectangular slab of concrete that is 15-by-20 square feet. She wants the shed to be 1 foot from each side of the slab. What dimensions will the shed have? What will its area be?" Tell the students to use number sentences in giving the solution to each of their problems—here, for instance, "The shed has to be 1 foot away from each side of the concrete slab, so I subtract 2 feet from the width of the slab (15 − 2), and 2 feet from the length of the slab (20 − 2), and then I know that the dimensions of the shed will be 13 by 18 square feet and that its area will be 234 square feet."

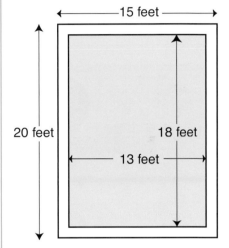

In the next activity, Changing Garden, students have an opportunity to explore what happens to perimeter and area measurements of a shape when the shape is modified in a particular way.

pp. 130–31

One-inch grid paper can be printed from the accompanying CD-ROM.

Changing Garden

Grades 4–5

Goals

- Calculate the perimeter of a rectangle
- Understand that for one given perimeter, many different areas are possible

Prior Knowledge

Students should have had experience in measuring area with informal and nonstandard units and in exploring the characteristics of rectangles.

Materials and Equipment

- A copy of the blackline master "Changing Garden" for each student
- Two or three pictures of gardens and garden fences from gardening magazines, books, or catalogs
- One sheet of one-inch grid paper for each student
- Twenty to thirty square-inch tiles for each pair of students

Classroom Environment

The students work in pairs on the activity and then come together as a class to discuss their work.

Activity

Engage

Show the class different pictures of gardens, and let the students share stories about gardens and gardening. Ask questions such as the following:

- "What types of gardens have you planted?"
- "What are some uses of gardens?"
- "What tasks are involved in planting a garden?"

Tell your students to suppose that they are designing a rectangular flower and vegetable garden that they could plant at school or at someone's house. Ask, "Do you think we might want to put a fence around our garden?" If your students say yes, ask them why. (Garden fences keep out animals that eat plants. Fences also prevent people and animals from running through gardens and accidentally stepping on plants.)

Pursue the subject: "What materials do people use to make garden fences?" After your students have made a few suggestions, tell them that people often use wire mesh to make fences around their gardens.

This activity is loosely based on ideas in Lappan et al. (1998a), pp. 35–36.

Explain that wire mesh is inexpensive and lets in sunlight, and some garden plants, like cucumbers, grow in vines that can climb on the wire. Ask the students to think about wire fencing for the garden that they are designing.

For more ideas about using a garden theme to develop students' measurement skills, see "Mathematics Instruction Developed from a Garden Theme" (Civil and Khan 2001) on the CD-ROM.

Explore

Tell the students to suppose that they have 30 feet of fencing to mark out and protect a rectangular garden. Remind the students that *perimeter* means the measurement around a figure, so the perimeter of this garden will be 30 feet. The students' task is to figure out the different rectangles that have a perimeter of 30 feet, given the following conditions:

- The fence must have no gaps or overlaps.
- The length of each side of the rectangular garden must have a whole-number measure.

Spend a few moments finding classroom objects with two-dimensional shapes that have noncurved sides. Try to include regular and irregular shapes as well as shapes that are concave and convex. Talk about the perimeters of these shapes.

Distribute the blackline master "Changing Garden." Group the students in pairs for the activity, and give each pair a sheet of one-inch grid paper and about thirty square tiles to use in finding all possible rectangles that fit the requirements. Say, "For this activity, let's agree that one inch on the grid paper and one inch on the side of a tile will represent one foot of fencing." Encourage the students to look for relationships between the perimeter (30 feet) and the length and width of each rectangle that satisfies the given conditions.

Before the pairs begin to work, review the difference between area and perimeter, as well as how to apply the formula for area of a rectangle. Ask the students to predict whether the areas of all the possible rectangular gardens will be the same or different. Students should record their predictions in their mathematics journals and then start on the investigation.

When they have finished, call the class together, and invite the students to share their findings. Work with an overhead projector and a transparency of the blackline master or draw a replica of the chart on the board. Fill in the chart with values offered by the students. Leave the area column blank for now.

Ask, "What discoveries did you make about the relationship between the perimeter of a rectangle and its length and width?" Students' answers might include comments like the following:

- "Add all the lengths and widths to find the perimeter."
- "First add the length and the width and then double the sum to get the perimeter."
- "Double the length, double the width, and then add these two numbers for the perimeter."

This question may prompt the students to generalize a method for finding the perimeter of any rectangle. If not, ask additional process questions:

- "Did you need to measure every side of the rectangle to find the perimeter? Why, or why not?"

- "How does knowing the characteristics of a rectangle help you find its perimeter?" (You need to measure only one length and one width, since a rectangle has two pairs of equal sides.)
- "If someone gave you the length and the width of any rectangle and asked you to find the perimeter, could you do it? If so, how?"

Once students have deduced the formula, help them write it algebraically in two forms: $P = 2l + 2w$ and $P = 2(l + w)$. Help students recognize that these two expressions are equivalent by expanding them and ordering terms to show both as $P = l + l + w + w$.

Next, complete the area column of the chart, and ask the students for their observations. Students should note that the areas are not all the same. In fact, some areas will be considerably larger than others. Encourage the students to extend their observations by considering how changes in the length and width of a rectangle change its area. Ask your students to look back at their predictions about the areas of different gardens. Were their predictions accurate? Why, or why not? Students should write their responses to this question in their journals.

Ask the students which rectangle they think would make the best garden for their school, and why. (Responses and reasons will vary. Students will probably choose the rectangle that makes the biggest garden and should be able to explain that this rectangle has a greater area than the others even though the perimeters of all the rectangles remain constant.)

Depending on the location of the school and the space available for a garden, students might decide that one of the other rectangles would be more desirable for a garden than the largest rectangle, even though it offers less garden space. Figure 3.4 shows a sample of students' work on the activity.

Fig. **3.4.**

Sample work by students on Changing Garden

1. Complete the chart below.

Fence	Length	Width	Perimeter	Area
A.	8	7	30 feet	50 sq. feet
B.	9	6	30 feet	54 sq. feet
C.	10	5	30 feet	50 sq. feet
D.	11	4	30 feet	44 sq. feet
E.	12	3	30 feet	36 sq. feet
F.	13	2	30 feet	26 sq. feet
G.	14	1	30 feet	14 sq. feet

2. In the space below, use words, pictures and /or numbers to describe how you found the perimeter of any size rectangle. To find a rectangle's perimeter you take all of the lengths and all of the widths and add them together to find your perimeter of your rectangle.

Extend

Have the students work together as a class to measure the perimeter of rectangular surfaces around the room. This time, encourage them to use fractions in expressing the dimensions. Students should first estimate the perimeter of an object and then measure the length and width and use the formula to find the actual perimeter.

Assessment Ideas

Ask your students to write two story problems—one related to area and the other related to perimeter. Give the students one condition for their problems: Nowhere may they use the word *area* or the word *perimeter*. Collect your students' problems and use them to create a page of story problems for everyone to solve.

Such exercises in writing and solving story problems related to perimeter and area will allow your students to demonstrate their ability to apply a specific strategy or formula. More important, these activities will provide students with opportunities to identify the attribute that they need to measure and the steps that they need to take when they are presented with a specific problem.

Where to Go Next in Instruction

After exploring real-life applications with a constant perimeter and a changing area, like the changing garden, students should also explore real-life applications with a constant area and a changing perimeter. Students should find all possible rectangular configurations of a given area with integer side lengths. They should calculate the perimeter of each rectangle and analyze the "look" of the rectangles that have the smallest and largest perimeters.

For example, students could find all the different ways to arrange 48 square tiles to make a rectangular kitchen floor. After finding the perimeter of each rectangle, students should explore the changes in perimeter as the dimensions change and discuss which rectangle might make the most useful kitchen floor. (Table 3.1 shows the possibilities.)

Table 3.1.
Possible dimensions and perimeters of a kitchen floor made with 48 square tiles

Floor	Length (Feet)	Width (Feet)	Area (Square Feet)	Perimeter (Feet)
A	48	1	48	98
B	24	2	48	52
C	16	3	48	38
D	12	4	48	32
E	8	6	48	28

Activities in grades 3–5 should address the development of the ideas underlying formulas for perimeter and area. The next two activities provide informal settings in which students can explore the area formula in the context of parallelograms and triangles.

Geo-Exploration– Parallelograms

p. 132

Geodot paper can be printed from the accompanying CD-ROM.

Grades 4–5

Goal

- Explore the formula *area = length × width* for the area of a rectangle and use it to develop the formula for the area of a parallelogram

Prior knowledge

Students should know the formula for the area of a rectangle (*area = length × width*) and have experience in using it.

Materials and Equipment

- A copy of the blackline master "Geo-Exploration—Parallelograms" for each student
- One geoboard for each pair of students
- Ten to twelve rubber bands for each pair of students
- A sheet of geodot paper for each student. (If geoboards are unavailable, geodot paper can be used alone instead of in combination with geoboards.)

For the teacher—

- Several sheets of geodot paper
- (Optional) An overhead geoboard
- (Optional) A transparency of the blackline master "Geo-Exploration—Parallelograms"

Classroom Environment

Students work on the activity in pairs but complete the activity pages and accompanying drawings individually. At the end of the activity, all the students come together as a class to discuss their work.

Activity

Engage

Use pictures or cutouts to review quadrilaterals with your students. Discuss the characteristics of squares, rectangles, parallelograms, rhombi, and trapezoids. Focus on the ways in which these shapes, particularly the rectangle and parallelogram, are similar to and different from one another. Be sure that your students recognize that the parallelogram is a two-dimensional shape with two pairs of parallel sides, and rectangles are the only parallelograms that have right angles.

Arrange your students in pairs, and tell them that they will be investigating the areas of parallelograms. Encourage them to continue to

This activity is loosely based on ideas in Barson (1971).

think about the similarities and differences between rectangles and parallelograms as they complete their explorations.

Explore

Distribute one geoboard to each pair of students, along with ten to twelve rubber bands. Give the students a few minutes to acquaint themselves with the geoboards, experimenting in any appropriate manner. Tell the students that they will use their geoboards and rubber bands to find a formula for the area of any parallelogram.

Review the meaning of the term *area* and the formula *area = length × width* (or $A = l \times w$) for the area of a rectangle. Say, "Let's agree to let the shortest distance between any two pegs on the geoboard be equal to one unit." Students should check the shortest distance and see that it is not the diagonal distance between pegs. Then ask the students to create five rectangles on their geoboards, record the dimensions, and find the areas with the formula $A = l \times w$. Have the students check their answers by counting the squares inside each rectangle. Be sure that the students are counting the distances between pegs, and not the pegs themselves, as units of length.

Next, tell your students to remove all the rubber bands from their geoboards except one, leaving just one rectangle. Ask them to transform this rectangle into another shape with the same area (see fig. 3.5). Say, "How can you be sure that this new figure has the same area as the rectangle?" Students should understand that the new shape could be reconfigured as the rectangle. They should verify this fact by counting the number of unit squares (including the sum of the partial squares) in the new shape to show that the number is the same as in the rectangle.

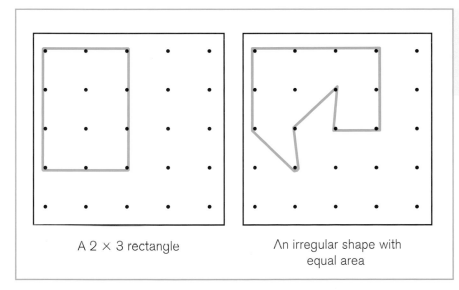

A 2 × 3 rectangle   An irregular shape with equal area

Fig. **3.5.**

A 2 × 3 rectangle and another shape (an irregular concave octagon) of equal area

 *The Geoboard applet on the CD-ROM lets students use virtual geoboards to investigate the area of parallelograms and triangles.*

Spend a few minutes discussing the terms *length*, *width*, *base*, and *height*. Be sure that your students understand that the length and width of a rectangle can also be described by the words *base* and *height*. Explain to the students that the height of a figure is the perpendicular distance from a vertex to the opposite side (the *base*). Tell the students that they can choose any side of a two-dimensional shape as its base.

Ask the students to work with their partners to make a parallelogram of any size on their geoboards. Have them identify and measure the base and height of their parallelograms.

Working with the whole class, extend the exploration of the base and height of a parallelogram. If an overhead geoboard is available, use it to consider several parallelograms. Otherwise, draw several parallelograms on sheets of geodot paper. Help students see that a parallelogram's height is the perpendicular distance from a vertex to the opposite side, which is its base. Look at different parallelograms and different vertices, to reinforce the idea that any side of a parallelogram can be its base.

Point out that, in the case of a rectangle, regardless of what side someone chooses to be the base, the rectangle's height will always be equal to one of the adjacent sides, since every angle in a rectangle is a right angle. Say to the students, "In a rectangle, the side that is *adjacent to*—that is, *next to* or *adjoining*—the base is the height."

Make sure that your students understand that this statement is not true for all two-dimensional shapes (see fig. 3.6). For an acute triangle, for example, no side can be equal to the height of the shape, since none of the triangle's sides is perpendicular to any of its other sides. Students should see that, in the case of a right triangle, if they choose one of the legs as the base, the other leg is the triangle's height, since the legs are perpendicular.

Fig. 3.6.

Heights (*h*) of selected two-dimensional shapes with base *b*.

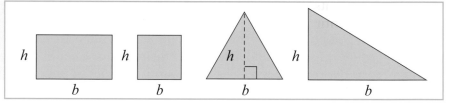

Ask, "Why isn't the side that is adjacent to the base of a parallelogram its height?" Be sure that your students understand that the height is perpendicular to the base. Since perpendicular lines form right angles and a parallelogram has right angles only if it is a rectangle, the height of a parallelogram will be one of its sides only in this special case. In all other instances, to "see" the height of the parallelogram, we must draw it (see fig. 3.7).

Fig. 3.7.

A rectangle (a) and a parallelogram (b) shown on geoboards. The height of the rectangle is the side, which is perpendicular to the base. The height of the parallelogram, as shown with a dotted line, is also perpendicular to the base but is not one of the parallelogram's sides.

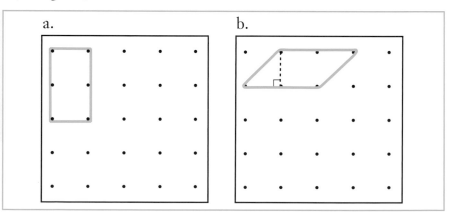

Your students should now be ready to explore the area of parallelograms. Give each student a copy of the blackline master "Geo-Exploration—Parallelograms" and a sheet of geodot paper. Again have the students work in pairs, this time to complete questions 1–4 on the blackline master. First, they will create five different parallelograms on their geoboards. They should find the length of each parallelogram's base and height and record them on the blackline master.

Finding the height of a figure without right angles can be challenging for students. Guide them in identifying the height of a parallelogram, stressing the idea of the height as the perpendicular distance from a vertex to the opposite side, which is the base (see fig. 3.8). Continue to review the definitions of *height* and *base* as well as of *perpendicular*, *opposite*, and *adjacent*.

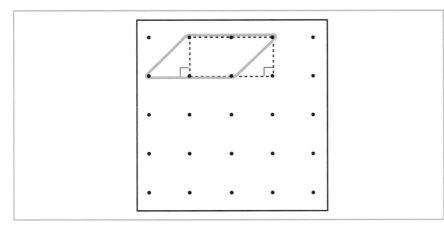

Fig. **3.8.**

A parallelogram with a rectangle of the same size superimposed

Give your students sufficient time to determine the area of each parallelogram by counting whole and partial squares. Let them record their data in the chart on the blackline master. Then bring the class together to discuss their work so far. If an overhead projector is available, make a transparency of the blackline master "Geo-Exploration— Parallelograms," and ask your students to help you fill in the chart in question 3. If no overhead projector is available, draw a chart on the board like the one shown.

Direct your students' attention to question 5 on the blackline master. This question asks them to look for patterns in their data and explain how they would find the area of any parallelogram. Let the students share their conjectures about a formula for the area of a parallelogram. They should see that they can find the area by multiplying the base times the height.

Look now at question 6, which asks the students to tell *why* the formula works. Let your students brainstorm about an explanation. One way to demonstrate the formula is to create a parallelogram on a geoboard with a rubber band, and use a second rubber band to make a rectangle that has the same base and height and is superimposed on the parallelogram so that the two figures share a leg. Figure 3.8 shows such an arrangement. Students should be able to see that this rectangle has the same base, height, and area as the parallelogram. In fact, they could cut out one shape, cut it into pieces, and rearrange the pieces to form the other shape. Thus, the formula for the area of a rectangle and the formula for the area of a parallelogram are the same.

Extend

Have students determine the missing dimension of a parallelogram when given the area and one linear measurement—either the base or the height.

Assessment Ideas

Ask students to make a poster titled "Area of a Parallelogram—How and Why?" (See fig. 3.9.) Students' posters should use words and pictures

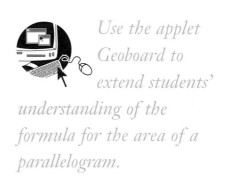

Use the applet Geoboard to extend students' understanding of the formula for the area of a parallelogram.

to explain the formula for the area of a parallelogram. The posters should include the formula $A = b \times h$ (in words or symbols), along with notes or pictures showing how to find and measure the base and height of a parallelogram, and an explanation using the similarities between rectangles and parallelograms to say why the formula works.

Fig. **3.9.**

A sample poster for "Area of a Parallelogram—How and Why?"

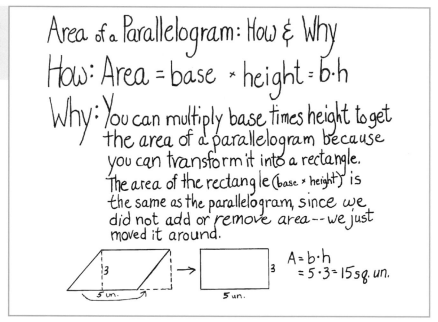

You can collect the posters and assess them individually. Later, you can also use them together with your students to make one class poster.

Where to Go Next in Instruction

Show your students drawings of a series of two-dimensional figures that are composed of rectangles and parallelograms. Figure 3.10 shows an example. Have students determine how and where to apply their formulas in order to find the total area of each figure. The area of the shape in figure 3.10 can be calculated by splitting it into a rectangle ($A = b \times h = 2 \times 4$) and a parallelogram ($A = b \times h = 1 \times 3$). The area of the rectangle is 8 square units. The area of the parallelogram is 3 square units. Thus, the area of the figure is 11 square units.

Fig. **3.10.**

A two-dimensional shape whose area can be measured by using the formulas for the area of a rectangle and the area of a parallelogram. The rectangle is 8 square units, and the parallelogram is 3 square units, for a total area of 11 square units.

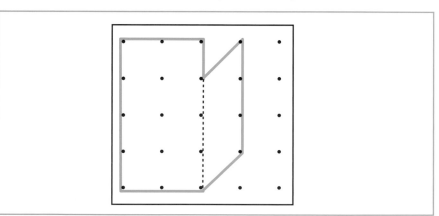

Geo-Exploration–Triangles

Grades 4–5

"All students should ... develop, understand, and use formulas to find the area of rectangles and related triangles and parallelograms."

Goal

- Develop the formula for the area of a triangle

Prior knowledge

Students should have had opportunities to identify the base and height of a parallelogram and to apply the formula for its area.

Materials and Equipment

- A copy of the blackline master "Geo-Exploration—Triangles" for each student
- One geoboard for each pair of students
- Ten to twelve rubber bands for each pair of students
- A sheet of geodot paper for each student. (If geoboards are unavailable, geodot paper can be used alone, instead of with geoboards.)

For the teacher—

- Several sheets of geodot paper
- (Optional) An overhead geoboard
- (Optional) A transparency of the blackline master "Geo-Exploration—Triangles"

p. 133

Classroom Environment

Students work on the activity in pairs but complete the activity pages and accompanying drawings individually. At the end of the activity, all the students come together as a class to discuss their work.

Activity

Engage

Pose the following problem: "Suppose that we are designing a triangular banner that we are going to hang in our classroom." Show the students a drawing of an isosceles triangle like that in figure 3.11. Make sure that you have specified dimensions as in the figure. Ask, "From what you know about the area of rectangles and parallelograms, how could you figure out the area of the banner?"

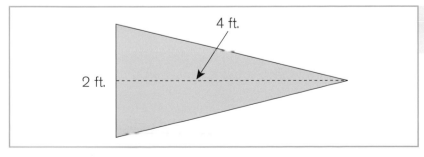

Fig. **3.11.**

Classroom banner

This activity is loosely based on ideas in Barson (1971).

(Alternatively, you might show your students a banner in the shape of a right triangle. Exploring the area of a right triangle can be easier than exploring the area of an isosceles triangle.)

Explore

Use a rubber band to make a triangle on a geoboard. (If an overhead geoboard is available, you can make a triangle on it.) Make a triangle that is similar to the one that you showed for the banner (an isosceles triangle if the banner is isosceles, a right triangle if the banner is a right triangle). Show your triangle to your students. Tell them that they are going to use the triangle as a model to solve the banner problem.

Be sure that your students can identify the base and height of the triangle. Discuss base and height and how they can be identified in acute, right, and obtuse triangles. On the chalkboard, draw an assortment of triangles in various orientations. Ask the students which side of each triangle they would like to serve as its base. Then, for each triangle, ask your students to identify or draw the height, making sure that it is perpendicular to the base.

Group your students in pairs and distribute to each pair a geoboard and ten to twelve rubber bands (or have the students use pencil and geodot paper if geoboards are unavailable). Give each student a copy of the blackline master "Geo-Exploration—Triangles" and a sheet of geodot paper, and ask the students to complete questions 1–4.

If some groups are having trouble finding a triangle's area, encourage them to compare the triangle to a parallelogram. (If they are working with a right triangle, they can compare it to a rectangle.). If students continue to struggle, show them how to superimpose a rubber band in the shape of a parallelogram over the triangle in such a way that the triangle and the parallelogram share a base and a side (see the illustration in the margin).

Give the students time to count squares (adding whole and partial squares) to find the area of this superimposed shape. As students work with their five triangles and the superimposed parallelograms, they will have an opportunity to recognize that the area of the triangle is half the area of the superimposed parallelogram. Since they know how to find the area of the parallelogram ($A = b \times h$), they can find the area of the triangle by making this calculation and dividing by two ($A = b \times h \times 1/2$).

If students create only right triangles, it is likely that they will discover the formula relatively quickly. If this happens, encourage them to explain why the formula works for the right triangles. Then ask them to consider whether the formula works for other types of triangles. Say, "Does your formula work for all types of triangles? Why, or why not?"

When the students have finished question 4 on the blackline master, bring the class together. If an overhead projector is available, make a transparency of the blackline master "Geo-Exploration—Triangles," and ask your students to help you fill in the chart in question 3. If no overhead projector is available, draw a chart on the board like the one in the blackline master.

Ask your students, "Has anyone discovered a formula for finding the area of a triangle?" If someone has, have him or her write the formula on the board, and then ask, "Why does this formula work?"

Students should have discovered the formula $A = 1/2 \times b \times h$. They should have some idea that the formula works because any triangle can

It is important for students to understand that base does not mean bottom and to recognize that the base of a figure is not always the side "on the bottom."

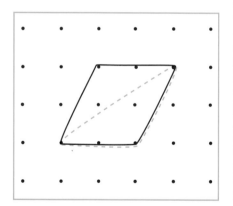

be considered as half of a parallelogram whose base and height are equal to the base and height of the original triangle. (See fig. 3.12.) By investigating the area of a triangle after having discovered how to find the area of a parallelogram, students can connect their new learning to previously learned concepts and make sense of the formula instead of merely memorizing it without understanding.

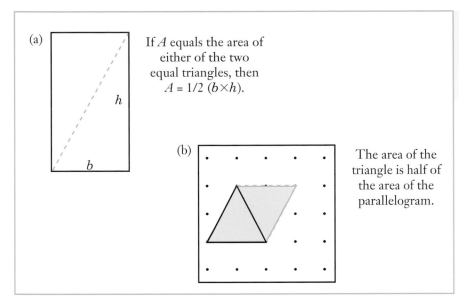

If A equals the area of either of the two equal triangles, then $A = 1/2\ (b \times h)$.

The area of the triangle is half of the area of the parallelogram.

Fig. **3.12.**

Determining the area formula for a triangle by (a) looking at a right triangle as half of a rectangle; (b) looking at any triangle as half of a parallelogram

Use the applet Geoboard on the accompanying CD-ROM to extend students' understanding of the formula for the area of a triangle.

Extend

Give students a variety of experiences in identifying and measuring the base and height of triangles and using these measurements to find area. Continue to ask students why the formula works. In particular, ask students to justify why it is necessary to take only half the product of the base and height.

Provide additional opportunities for students to explore the formula on the geoboard. While students are still developing their understanding of the formula, they should continue to superimpose rubber-band parallelograms over the triangles to see the relationship between the two.

Assessment Ideas

To assess your students' learning from this activity, ask them to reconsider the banner problem. Have the students choose dimensions for a different triangular banner and find its area. Students should show how they found the area by using what they know about the area of rectangles and parallelograms.

Where to Go Next in Instruction

Principles and Standards for School Mathematics recommends that students in grades 3–5 find formulas for the area of a rectangle, a parallelogram, and a triangle. Your students will need many opportunities to apply the three formulas in real-life problem solving. These investigations will accustom them to finding the area and perimeter of a rectangle as well as the area of a parallelogram and a triangle.

Your students should also be able to think of a variety of situations in which they would need to find the area or perimeter of an object. For each, they should consider the unit that is most suitable for the measurement.

In addition, students in grades 3–5 should be able to interpret a measurement story problem, deciding what attribute needs to be measured to solve the problem. Do they need to determine the area of a shape or its perimeter?

After many such investigations, students will be ready to consider how to measure the surface area and volume of rectangular three-dimensional shapes. It is important for students who are learning these ideas to make connections between two-dimensional and three-dimensional measurements. For example, the formula for the volume of a rectangular prism (*length × width × height*) rests on finding the area of the base of the prism with the formula for the area of a rectangle (*length × width*).

Instruction that highlights these connections is likely to promote a strong understanding of the process and applications of measurement. The following chapter extends some of these topics.

NAVIGATING *through* MEASUREMENT

Chapter 4
Measuring Three-Dimensional Objects

Students in grades 3–5 need to explore rectangular solids, investigating and measuring their volumes and surface areas, as well as their weights and capacities. This chapter features four activities that help students select and use benchmarks to predict such measurements, determine appropriate standard units for them, and develop a variety of strategies that lead to generalizations or formulas. The Measurement Standard elaborated in *Principles and Standards for School Mathematics* (NCTM 2000) emphasizes these skills.

The chapter opens with two activities that help students in grades 3–5 meet the Measurement Standard's expectation that they will "develop strategies to determine the surface areas and volumes of rectangular solids" (NCTM 2000, p. 170). The first activity, Building Boxes, focuses on finding the volume of a rectangular solid. Students use their understanding of multiplication to predict and estimate the space inside a variety of rectangular solids, and then they use cubes to measure the space, ultimately generalizing a formula. The second activity, Wrap It Up! invites students to explore surface area by covering or wrapping a rectangular solid. As students construct different-sized boxes, they arrive at the general idea that surface area is the sum of the measures of the area of each side of a solid.

The third activity, Sizing Up Funny Shapes, allows students to explore the volumes of irregularly shaped solids, experimenting to "develop strategies for estimating the … volumes of irregular shapes" (NCTM

2000, p. 170), as the Measurement Standard recommends. The activity introduces students to displacement as a technique for measuring the volume of irregular solids.

The final activity, Weighing In at the Carnival, reinforces ideas about units of weight in both the metric and the customary systems. Students become familiar with the standard units kilogram and pound and use them as benchmarks to predict the weights of objects before actually measuring them. *Principles and Standards* emphasizes the importance of developing such benchmarks as lifelong tools.

Throughout the chapter, students use measurement of three-dimensional objects in everyday situations. Such real-life contexts are indispensable to their seeing an immediate connection between the mathematics of measurement and routine activities in their daily lives.

Building Boxes

Grades 3–5

Goals

- Predict how many cubes will fill a box
- Check the prediction by building the box and filling it with cubes
- (By grade 5) Develop strategies for calculating the number of cubes needed to fill a box
- Compare the numbers of cubes needed to fill two boxes

Prior Knowledge

Students should be acquainted with rectangular prisms.

Materials and Equipment

- A copy of the blackline master "Building Boxes" for each student ("Building Boxes—A" is for students in grades 3; "Building Boxes—B" is for students in grade 4 or 5)

For each pair of students—

- Five sheets of two-centimeter grid paper
- About 65 two-centimeter cubes. (If two-centimeter cubes are unavailable, be sure that whatever cubes and grid paper you choose match in scale.)
- One roll of tape
- One pair of scissors

For teachers of grade 3—

- A transparency of the blackline master "Building Boxes—A"

Classroom Environment

Students work in pairs.

Activity

Engage

Say to your students, "Suppose you are package designers for the Box-M-Up Package Company." Ask the students to imagine that their department designs and constructs boxes without lids. Workers in a different department measure the boxes and make lids.

Group the students in pairs, and give each pair five sheets of two-centimeter grid paper. Distribute copies of the blackline master "Building Boxes" to each student. (Use "Building Boxes—A" with third graders and "Building Boxes—B" with fourth or fifth graders.) Explain that, working in pairs, the students are going to use the grid paper to make scale models of boxes for their company. Take five sheets of grid paper for your own use in demonstrating the process.

Hold up one of your sheets of grid paper and trim it so that you are left with a rectangle that is 9 squares by 11 squares. Direct your students to trim each of their five sheets of grid paper to these dimensions, taking

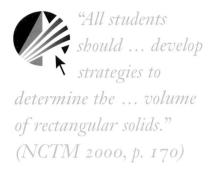

p. 134, 135

The CD-ROM includes a template for two-centimeter grid paper. Printing on paper of assorted colors will enable you to give each pair of students sheets of grid paper for boxes in distinctive colors.

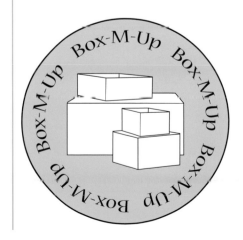

care not to cut into the squares. Trim all of your remaining sheets of grid paper to the same size.

If you are teaching third graders, put a transparency of "Building Boxes—A" on an overhead projector (or draw a replica of the chart on the board). In the first row of the chart, record the length and width of the trimmed grid paper. Direct your students to do the same on their copies of the blackline master.

Also in the first row, enter zeros in the columns labeled "Height (number of layers of unit cubes)," "Volume (number of unit cubes needed to fill box)," and "Total number of unit squares cut from rectangle." Explain that the students will be cutting squares from the corners of their trimmed grid-paper rectangles in making boxes, but the flat trimmed grid paper is not a box, so it has no height or volume.

If you are teaching fourth or fifth graders, your students should be able to fill in the chart on the blackline master "Building Boxes—B" in collaboration with their partners without special guidance from you, and then they can begin on their own to discuss any patterns that they see in their data.

Ask the students to look at their trimmed sheets of grid paper and tell how many squares they have on each trimmed sheet. Before demonstrating how to construct the first box, make sure that the students agree that each sheet has 99 unit squares.

Hold up a sheet of trimmed grid paper, and cut one square from each corner, as shown in figure 4.1. Fold up the outside rows to make a box, taping the corners. Show your box to your students, and direct them to follow the same process to make their first box. Then ask them to fill the box with two-centimeter cubes and report to the class how many cubes they needed to fill their box.

Fig. **4.1.**

A unit square cut from each corner of a 9-by-11-unit rectangle to make a box with a height equal to one unit

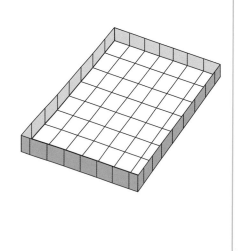

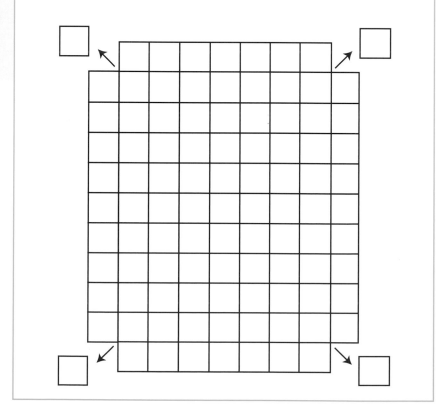

As the pairs are constructing this first box and filling it, observe what they do, and listen to what they say. What strategies do they use to count the cubes? Do they need to fill the entire box before they know how many cubes fill it?

Once all the pairs have completed the task, bring the class together and let each pair report its findings. Ask the students how they arrived at the number of cubes to fill the box. Did they count every cube? Did they count the number of cubes in the top row and the number of rows and multiply? When the students agree that 63 cubes are needed to fill the box, have them share the strategies that they used to arrive at that total.

If you are teaching third grade, call your students' attention again to the transparency of "Building Boxes—A" (or your replica of the chart on the board), and elicit their help as you enter information for box 1 in row 2. Let the students fill in or correct the information on their own copies of the blackline master.

Explore

Return the students to their pairs, and say, "Now let's design a new box that your team will construct." Hold up your second sheet of trimmed graph paper, and cut a 2-by-2-unit square from a corner. (See fig. 4.2.) Explain that students will cut a square of this size from each corner of their second 9-by-11-unit rectangle to construct the next box. Ask students to predict whether they will need more or fewer than 63 cubes to fill this new box.

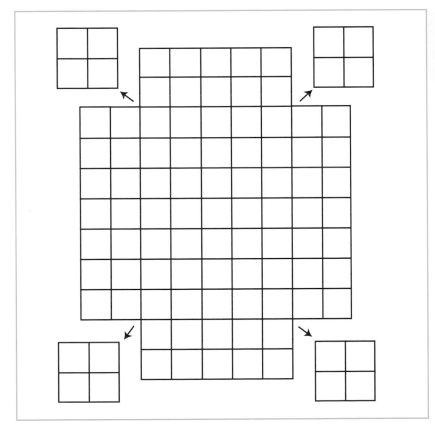

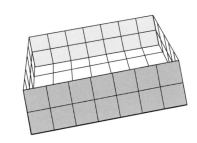

Fig. **4.2.**

A 2-by-2-unit square cut from each corner of a 9-by-11-unit rectangle to make a box with a height equal to 2 units

Have your students register these predictions by a show of hands, and write the results of your informal poll on the board. Students may imagine that this second box will hold fewer cubes than the first. They may think that it is smaller than the first one, especially if they do not consider the effect that its greater height has on its capacity. Do not prompt your students to consider the change in height from the first box to the second.

Direct the students to construct the second box. Tell them to let you know if and when they want to change their predictions.

Again observe what the students do and say as they construct and fill the box. Note any strategies that they use in determining the total number of cubes needed to fill it. See if the students have to fill the entire box to arrive at the total.

When all the pairs have completed this new task, bring the class together again to share findings. If your students are third graders, again enter the information on the transparency (or on the board). When all the students agree that the second box holds 70 cubes, let them discuss their predictions, comparing them with their results. Were their predictions right, or were they wrong? If they were wrong, what aspects of the situation did the students overlook? If a student predicted accurately, what aspects of the situation did he or she consider in making the prediction?

At the end of this discussion, say, "Let's design another box." Let your students watch again as you cut a 3-by-3-unit square from a corner of your third sheet of trimmed grid paper. Again ask the students to estimate the number of cubes that the box will hold.

Students may think that because the second box held more than the first, this third box will hold yet more than the second. Direct them to construct the third box, again letting you know if and when they want to change their predictions.

As before, observe what the students do and say as they construct and fill the box, noting any strategies that they use to determine the total number of cubes. Do they need to fill the entire box to arrive at the total this time?

When all the pairs have completed this task, bring the class together once more to share findings. Third graders again should watch you enter the correct information into the chart on the transparency (or on the board).

After the students agree that the third box holds 45 cubes, ask them to build a fourth box. Again demonstrate how they will make the new box, by cutting a 4-by-4-unit square from a corner of their fourth rectangle. Let the students follow the same process of predicting and constructing the box to confirm or disprove their predictions.

When the students agree that the fourth box holds 12 cubes, ask them if they can use the same process to make a fifth box. Watch what they do with their fifth rectangle. Actually, the rectangle is too small for them to produce another box in the same way, but let your students discover this fact for themselves.

After making the boxes and completing the chart, students—especially fifth graders—should have begun to see that length times width gives them the bottom "layer" of their box, and when they multiply by the number of layers (height), they find the total number of cubes that the box will hold. The process of making the boxes and filling them should have helped all students in grades 3–5 develop a mental construct that makes the concept of volume easy to grasp.

Define *volume* as the space that a three-dimensional object occupies. Stress that the students have been measuring the space that the boxes occupy—their volume—by the numbers of cubes needed to fill them.

If you are teaching grade 3, have your students consider questions 1 and 2 on "Building Boxes—A." These questions are intended to elicit their ideas about what happened to length, width, and volume with each new box and how they figured out how many cubes would fit in a box.

If you are teaching grade 4 or 5, ask your students to complete question 1 on "Building Boxes—B." Here they describe all the patterns that they can find in the data in the chart.

Discussing your students' responses to these questions can help you avoid teaching the formula (*volume = length × width × height*) directly and having your students learn it by rote. If students arrive at this formula, help them see that although it works with rectangular solids, like their boxes, it would not help them find the volume of a sphere, for example. Students should identify a number of patterns:

- The length decreases by 2 as each corner is cut.
- The width also decreases by 2 as each corner is cut.
- The height (number of layers) increases by 1 as each corner is cut.
- The number of cubes needed to fill a box is the number in the bottom layer, which the students may be able to describe as *length* times *width*, times the number of layers, which they may be able to describe as *height*.

Extend

To extend the activity, you might propose that your students work on another box-making project for the Box-M-Up Package Company. Give the student pairs additional sheets of two-centimeter grid paper, and suggest that they begin with a square grid this time—say, 9 by 9 units or smaller, depending on how many boxes you want your students to make. Ask—

- "How many different boxes with whole-number sides can you make now?" (See table 4.1.)
- "Which box holds the most cubes?"
- "List and explain all patterns that you see."
- "How do these patterns differ from those that you found in the first activity?" (Here all layers of the boxes are squares.)

Table 4.1

Volumes and Surface Areas of Boxes Made from a 9-by-9-Unit Square

	Length	Width	Height (Number of Layers of Cubes)	Volume (Number of Cubes Needed to Fill Box)	Total Number of Squares Cut from Rectangle	Number Squares in Grid-Paper Net
Paper square	9	9	0	0	0	81 squares
Box 1	7	7	1	49 cubes	4 squares	77 squares
Box 2	5	5	2	50 cubes	16 squares	65 squares
Box 3	3	3	3	27 cubes	36 squares	45 squares
Box 4	1	1	4	4 cubes	64 squares	17 squares

NCTM's Illuminations
Web site provides related
activities that can extend or
be used in assessing
students' learning from the
activity Building Boxes. See
http://illuminations
.nctm.org/reflections/3-5/
GatheringEvidence/
goals.html.

Students in grades 4 and 5 also should begin to realize that it is not always possible or convenient to fill a box with unit cubes to find its volume. Their investigations should point them toward a generalization about the volume of any rectangular solid. They should discover that they can multiply length times width to find the number of cubes in the bottom layer (the area of the base) and then multiply the result by the number of layers (the height) to find the volume of a box. When they reach this conclusion, write the formula for the volume of any rectangular solid on the board: *volume = length × width × height*, or *volume = (area of the base) × height*.

Assessment Ideas

Noticing what your students do and say as they make the boxes and complete the chart will help you assess their understanding in this investigation. After you model the process of organizing the data in the chart, pay particular attention to the patterns that students can identify in writing.

Students in grades 4 and 5 can also demonstrate their understanding by writing in their mathematics journals. You can assess their learning by examining their answers to such questions as these:

- "Could you figure out how many cubes would fit in a box without actually filling it?" (Students may suggest using the formula for the volume of a rectangular prism.)
- "What would be the dimensions (length, width, and height) of a box that you could build from a 10 by 10 rectangle by cutting one square from each corner of the paper?" (length = 8; width = 8; height = 1)
- "Draw a picture of this box, showing the relationships among length, width, and height as accurately as you can."
- "If you cut a 3-by-3-unit square from each corner of an 8-by-12-unit rectangle, how many unit layers would the resulting box have?" (3)
- "How do you know that you would have this many layers?"
- "How many cubes would fit on the bottom layer?" ($2 \times 6 = 12$)
- "Draw a picture of this box, showing the relationships among length, width, and height as accurately as you can."

Where to Go Next in Instruction

Throughout the year, return to this activity periodically, giving students different numbers of square units and asking them to design all the possible boxes and compare their volumes. They should make charts for these activities like the chart in the "Building Boxes" blackline master.

For students in grades 4 and 5, the next activity, Wrap It Up! can take students another step forward by helping them link the ideas of volume and surface area.

Wrap It Up!

Grades 4–5

Goals

- Develop strategies to determine the surface area of rectangular prisms
- Distinguish between the volume and the surface area of three-dimensional objects
- Use nets to relate the surface areas of 3-D objects to the areas of 2-D shapes
- Determine the smallest and largest possible surface areas of a rectangular prism with a fixed volume

Materials and Equipment

For each pair of students—

- Copies of the blackline masters "Researching Box Designs" and "Researching Box Designs—Extension"
- A copy of "Box Net 1" and "Box Net 2" from the CD-ROM
- A set of 18 two-centimeter cubes
- Eight to ten sheets of two-centimeter grid paper (possibly in assorted colors, for distinctive boxes)
- Scissors and a roll of tape
- A sheet of tag board or poster board
- Markers or crayons
- (Optional) Decorative materials (stickers, glitter, etc.)

Prior Knowledge

Students should know that the dimensions of a rectangular prism are its *length*, *width*, and *height*, and their vocabulary for describing the prism should also include *face*, *edge*, *base*, and *vertex* (see fig. 4.3). They should have made nets of rectangular prisms, and they should have explored the *perimeter* and area of 2-dimensional shapes and the *volume* of 3-dimensional shapes.

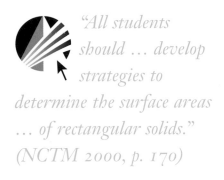

"All students should … develop strategies to determine the surface areas … of rectangular solids." (NCTM 2000, p. 170)

pp. 136–37, 138

If you make two-centimeter grid paper for Wrap It Up! from the template on the CD-ROM, your students will need to tape trimmed sheets together to have a grid large enough for a box whose dimensions are 12 × 1 × 1.

Surfaces of rectangular prisms are *faces*.
The top and bottom faces are *bases*.
The intersection of two faces is an *edge*.
The identified edge represents the *height* of the prism.
The intersection of three edges is a *vertex*.

Fig. **4.3.**

A rectangular prism

See Navigating through Geometry in Grades 3–5 (Gavin et al. 2001) for additional geometric terms and concepts for students in grades 3–5.

This activity has been adapted with permission from Lappan et al. (1998b), p. 16.

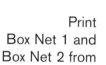
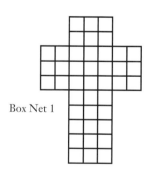

Box Net 1

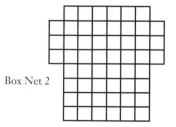

Box Net 2

Classroom Environment

Students work in pairs.

Activity

Engage

Tell your students to imagine that they are box designers for the Ooey-Gooey Candy Company. Ooey-Gooey makes caramels and packages them twelve to a box in rectangular boxes that are then sold in retail stores. Hold up some two-centimeter (or other-sized) cubes, and say, "The caramels have the same dimensions as these cubes."

Don't give the metric dimensions of the cubes; this information could easily confuse the students as they work. Instead, make sure that you talk about the cubes as measuring 1 unit on a side. Thus, the volume of a box of caramels will be fixed at 12 cubic units, and its surface area will be a whole number of square units.

Explain that the Ooey-Gooey Candy Company wants boxes in sizes that are appealing to customers, easy to ship, and convenient to stack and store. Of course, the company is seeking to make a profit, as well.

Say to your students, "Suppose your job is to review all the options for rectangular boxes and prepare a presentation in which you will recommend a specific box to Ooey-Gooey's board of directors." Tell the students that their presentations have to include a model of the box that they are recommending as well as their reasons for their choice.

Print Box Nets 1 and 2 (shown in the margin) from the templates on the CD-ROM. These nets are for boxes of equal volume (though not 12 cubic units). Use them as a bridge from your students' experiences in the activity Building Boxes. Cut out the nets. Fold them and tape them to make boxes with lids. Tell your students that the Ooey-Gooey Candy Company, unlike the Box-M-Up Package Company, manufactures each box and lid in this way—all in one piece.

Ask your students, "How are these two boxes alike?" When they say that the boxes hold the same amount (in other words, their volume is the same), change the focus from similarities to differences, and ask, "How are they different?" Students may say that the boxes have different shapes. If so, ask them to expand on this idea. (They might say that one looks more like a cube than the other; one is longer than it is wide, etc.).

To continue to explore the solids, ask your students a question that calls for exact numbers: "What are the dimensions of each box?" (The box from net 1 is $3 \times 3 \times 2$ cubic units, and the box from net 2 is $6 \times 3 \times 1$ cubic units.) Turn the students' attention to the nets of the boxes by asking the following question: "If we cut the tape and unfolded these boxes, what would we have?"

You may need to define *net* for your students. A net is a two-dimensional pattern that can be folded to make a three-dimensional model of a solid, such as a box. Teachers of children in prekindergarten through grade 2 sometimes describe such a pattern for a solid as a "jacket," which is not a bad image.

Show the nets of the two boxes to your students by cutting the tape and unfolding the boxes. Ask, "How much area does each net cover?" (Net 1 covers 42 square units; net 2 covers 54 square units.) Pose the following question for your students to think about as they to work on the problem of the candy boxes: "How does the area of the net relate to the box that you can make?" As the students work, they should realize that the amount of material required to construct a box is equal to the area of the net.

Explore

Give each pair of students twelve cubes, several sheets of grid paper, scissors, tape, tag board or poster board, markers, and a copy of the blackline master "Researching Box Designs." Say, "In your report, you must show that you have explored all the options and state your recommendation. You should explain the reasons for your choice very clearly. Your presentation should include a 3-dimensional model of the box that you are recommending."

The students will then use the grid paper and cubes to discover all the possible ways to make boxes for 12 caramels. They should assume that each box holds exactly 12 candies with no unoccupied space. Students should discover that there are four different combinations of dimensions for boxes for 12 caramels—a box that is $12 \times 1 \times 1$ (or $1 \times 12 \times 1$ or $1 \times 1 \times 12$), a box that is $6 \times 2 \times 1$ (or any other permutation of these numbers), a box that is $4 \times 3 \times 1$ (or any other permutation), and a box that is $3 \times 2 \times 2$ (or any other permutation).

For each possible box, the students should make a net, cut it out, fold it, and tape it together. (They should ignore the need for flaps to keep box lids closed.)

If necessary, help your students fill out the chart on the blackline master so that they can organize their findings and recognize patterns (see fig. 4.4). The data in the chart may show them that the number of different boxes is related to the factorizations of 12.

Fig. **4.4.**

Sample completed chart

Dimensions of the Box	Volume of the Box	Sketch of a Net of the Box (One is shown for each; however, there are others.)	Surface Area of the Box
$12 \times 1 \times 1$	12 cubic units		50 square units
$6 \times 2 \times 1$	12 cubic units		40 square units
$4 \times 3 \times 1$	12 cubic units		38 square units
$3 \times 2 \times 2$	12 cubic units		32 unit squares

When all your pairs are ready, summarize the activity by gathering the data that the students have collected on the blackline master. This is a good opportunity for you to model the process of collecting and recording data in an organized manner. Start with the 12 × 1 × 1 box and fill in the volume of the box (12 cubes) and the area of the net (50 squares). Then ask, "Did anyone find another box that has a height of 1 with different lengths and widths for the base?" Continue until all boxes with heights of 1 have been listed. Move to a height of 2, and then ask if a height of 3 is possible.

This process will inevitably raise an important question: Are two boxes with dimensions like 4 × 1 × 3 and 4 × 3 × 1 the same? (See fig. 4.5.) This discussion will provide you with an excellent opportunity to explore spatial *orientation* with your students. Consider with them whether or not orientation makes a difference for boxes of Ooey-Gooey caramels.

The orientation of a box of caramels could be an important consideration to a retail store in allocating space, stacking the boxes, and displaying them on a shelf. However, if the volumes and surface areas of differently oriented boxes are the same, they should not be considered different for this activity or for Ooey-Gooey's box-manufacturing process.

Fig. **4.5.**

A 3 × 4 × 1 box in two different orientations

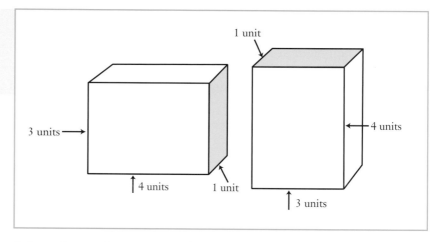

Ask students what patterns they see in the data in their charts. They might say that the volume of a box here is always 12 cubes. They might discover that the product of the length, width, and height of a box is always equal to the volume. They might observe that the only numbers possible for the dimensions of Ooey-Gooey's boxes are the factors of 12. Let their mathematical discoveries lead the discussion.

Be sure to emphasize the kind of units used to represent surface area (square units) and the kind used to represent volume (cubic units). Ask, "Why do we use square units for area?" The net offers an excellent visual representation of the squares that we use in measuring area. In showing your students a net, you can stress the fact that surface area represents the number of squares that cover the surface of the box. Turning to volume, ask, "Why do we use cubic units for volume?" (Volume is a measure of three-dimensional space, and in the activity it represents the number of cubes that fill up the space inside a box.)

Next, have the pairs of students make their presentations to the class, showing their models and any sketches of nets that they may have used to determine surface area. Make sure that the students explain why they chose their particular design. They can give a variety of reasons for their choice, such as that it will be appealing to customers or will be easy to stack on shelves. You may have at least one group that focuses on surface area and chooses the $3 \times 2 \times 2$ box as the most cost-effective design, since it uses the least material.

To bring the activity to a close, help students articulate the mathematical concepts that they have learned about volume and surface area. To assist this process, ask—

- "What do we call the measure of the number of cubic units that a box will hold?" (Volume)
- "What is the volume of the boxes that you were working with for the Ooooey-Gooey Candy Company?" (12 cubic units)
- "Did the volume change from box to box?" (No)

Turn to surface area. Say to your students, "The area of the net that you made for each box represents the surface area of the box, or the amount of material that you would need to make or cover the box." Probe your students' understanding by asking—

- "Can you give me a general rule for finding the surface area of any box?" (It is the sum of the areas of each face of the box.)
- "We said the volume stayed the same for all the boxes, but what happened to the surface area?" (The surface area changed from box to box, depending on the dimensions.)
- "When might you want to find the surface area of something?" (Students might suggest that they could use surface area to determine the amount of wrapping paper that they would need for a gift, to determine the amount of foil or plastic wrap that they would need to wrap food, etc.)

Extend

Have your students take another look at the boxes that they made for the Ooey-Gooey Candy Company and the data that they organized in their charts. Challenge them to consider the shapes of the boxes in relation to their surface areas. Ask, "Which box has the least surface area?" (The design that is $3 \times 2 \times 2$). Have them think closely about the shape of this box: Ask, "What could you say in describing this box?" (Students should arrive at the idea that the box's dimensions are all quite similar—that is, they are "close" to one another in size. Of all the possible boxes, this one looks most like a cube.)

Next, ask, "Which box has the greatest surface area?" Urge them to think closely about this box by asking, "How could you describe it?" (It is long and narrow; its dimensions show the greatest difference in size.)

Give each pair of students six more cubes, additional sheets of grid paper, and a copy of the blackline master "Researching Box Designs—Extension." Tell them to imagine that the Ooey-Gooey Candy Company has changed its marketing plans and now needs them to design a box that will hold 18 caramels. Have them use the blackline master to organize their data as they investigate the following questions:

- What are the dimensions of the box that has the greatest surface area?" ($18 \times 1 \times 1$)
- "Why is this true?" (The cubes are spread out in a way that makes the largest possible surface for a box to cover.)

- "What is the surface area of this box?" (74 square units).
- "What are the dimensions of the box with the least surface area?" (3 × 3 × 2).
- "What is the surface area of this box?" (42 square units)

You could have your students continue their explorations with 11, 16, 24, and 27 cubes. Then you could say, "Suppose someone tells you what the volume of a box has to be." Make sure that the students are thinking of a box whose volume and dimensions are whole numbers of units. Ask them to consider the following questions:

- "Can you find the box with this volume that has the greatest possible surface area?"
- "Can you find the box with this volume that has the least surface area?"
- "Can you describe the shape of each box?"

The closer the shape of the box is to a cube, the smaller its surface area. The box with the maximum surface area will be long and "skinny" and have two dimensions that are 1 unit.

Assessment Ideas

During the activity, you can assess your students' understanding of volume and surface area informally by listening to the pairs of students as they construct their nets and boxes and complete their charts. When students present their reports and recommendations to the class, be sure that they explain how they constructed the box, give its volume and surface area, discuss how the net relates to the box, and explain why they chose the box that they did.

You could also use an individual writing assignment to assess your students' learning. You could ask your students to write letters to the president of the Ooey-Gooey Candy Company, outlining their recommendation and enclosing a model and a sketch of the net for the box of their choice. Remind them to discuss the box's volume and surface area and to state their reasons for their recommendations as clearly and convincingly as possible.

Where to Go Next in Instruction

Elementary school students will need additional practice to consolidate their newly acquired ideas about surface area and volume. Throughout the year, refresh students' understanding periodically by giving them different numbers of cubic units and asking them to create all the possible rectangular prisms, find a net for each one, think about surface area of each in relation to its volume, and compare surface areas of all the prisms by completing and examining charts like those in the blackline masters for this activity.

After your students have some measuring experiences with the volume of solids through activities such as Building Boxes and Wrap It Up! the next logical step is for them to explore displacement as a strategy for measuring the volume of irregular solids, such as rocks and pens. The next activity, Sizing Up Funny Shapes, provides students with such an opportunity and familiarizes them with metric measures of weight.

Sizing Up Funny Shapes

Grades 3–4

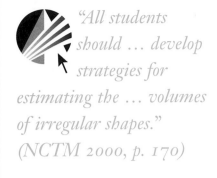

"All students should … develop strategies for estimating the … volumes of irregular shapes." (NCTM 2000, p. 170)

Goals

- Measure volume of common objects by building models with centimeter cubes
- Measure volume in cubic centimeters by water displacement
- Read calibrated measuring devices
- Understand that all measurements are approximations

Prior Knowledge

Students should have explored the volume of rectangular prisms and have had opportunities to use metric measurements for volume. (This activity could follow Building Boxes.)

pp. 139–40

Materials and Equipment

- A copy of the blackline master "Sizing Up Funny Shapes" for each pair of students

For each group of four students—

- Twenty to thirty one-centimeter cubes (connecting or nonconnecting)
- A set of metrically calibrated measuring tools, such as graduated cylinders or measuring cups
- A container of water or uncooked rice
- Small, irregularly shaped objects that will not be damaged by water. (Sample objects include a crayon, a marble, a lump of modeling clay, a rock from the playground, and an orange)

For the teacher—

- A glass or clear plastic pitcher
- A small assortment of pebbles or marbles
- Containers measuring one quart and one liter

Classroom Environment

Students work at stations that are set up to accommodate four students at a time. The students work in pairs within each group of four.

Activity

Engage

Read aloud to your students "The Crow and the Pitcher" from Aesop's Fables (see fig. 4.6). Then ask the students to explain in their own words what happened in the story.

This activity has been adapted with permission from the Teaching Integrated Mathematics and Science (TIMS) Project (1997), pp. 218–20.

Fig. **4.6.**

Fig. 4.6. Aesop's fable of the Crow and the Pitcher (from Aesop's Fables Online Collection at http://www.aesopfables.com/)

The Crow and the Pitcher

A Crow, half-dead with thirst, came upon a Pitcher which had once been full of water; but when the Crow put its beak into the mouth of the Pitcher he found that only very little water was left in it, and that he could not reach far enough down to get at it.

He tried, and he tried, but at last had to give up in despair. Then a thought came to him, and he took a pebble and dropped it into the Pitcher.

Then he took another pebble and dropped it into the Pitcher.

Then he took another pebble and dropped that into the Pitcher.

Then he took another pebble and dropped that into the Pitcher.

Then he took another pebble and dropped that into the Pitcher.

Then he took another pebble and dropped that into the Pitcher.

At last, at last, he saw the water mount up near him, and after casting in a few more pebbles he was able to quench his thirst and save his life.

Little by little does the trick.

Fill a glass or pitcher about halfway with water. Drop pebbles or marbles (or centimeter cubes) into the water one at a time, and have your students describe what happens after you drop each object. Allow students to drop some of the objects into the water, too.

Explore

Explain to your students that in the metric system liquids are measured in liters (and fractions and multiples of a liter), and tell them that a liter is slightly larger than a quart (1 liter = 1.06 quart). Show the students containers that hold one quart and one liter. Put the quart aside while continuing to display the liter. Say, "If this is a liter, what do you suppose a milliliter is?" If necessary, explain to your students that *milli-* is a prefix meaning thousandth. Ask them to show you how big they think a milliliter would be in relation to your one-liter container.

After the students have exchanged ideas, tell them that a milliliter is equal to 1 cubic centimeter. Students can easily explore this equivalence by placing centimeter cubes into a graduated cylinder calibrated in milliliters and observing that the water rises 1 milliliter for each cube that they place in the cylinder.

Explain to your students that *displacement* is what we call this strategy of measuring an object's volume by submerging it in fluid and measuring the change in the level of the fluid in its container. The water rises in the cylinder because is has been "pushed aside," or *displaced*, by the cube. Use this opportunity to demonstrate how to read a graduated cylinder correctly: To get an accurate reading of the water level, students must position themselves so that their eyes are level with the liquid in the cylinder.

For the next part of the activity, you'll need to have your classroom organized into displacement stations at which four students can work at a time. To set up the stations, select some common irregularly shaped classroom objects, taking care to choose items that will not float or be damaged in water. Some possible objects are identified in the materials list, and others might include a plastic domino and a ball that does not float. Along with a small collection of such objects, each displacement station will need a set of metrically calibrated measuring tools, twenty to thirty centimeter cubes, and a large container of water.

For the displacement portion of the activity, your students can use rice instead of water, if you prefer. To measure with rice, they should partially fill a large calibrated measuring tool, such as a measuring cup or graduated cylinder, with rice. Then they can follow the same process as with water, except that they must push the object to be measured into the rice until it is completely covered. They will determine the object's volume just as in water; the difference between the levels of the rice before and after the object has been buried will represent the volume of the irregularly shaped object. Using rice instead of water permits students to obtain approximate measures for the volumes of irregularly shaped objects that might be damaged by water.

Divide your students into pairs, and give each pair a copy of the blackline master "Sizing Up Funny Shapes." Explain that each pair will work alongside another pair at one of the displacement stations in the room. Arrange the pairs in groups of four, and send each group to a displacement station.

Step the students through the process that they will be following with each object that they measure at the displacement station. First, have each pair select an irregularly shaped object to measure. Then tell them that they must try to make a replica of their object with the one-centimeter cubes that they find at the station. Have them count the cubes in their replica, and explain that the total will give them an estimate of the number of cubic centimeters in the volume of their object. (If their object is too large for them to replicate it with their own supply of centimeter cubes, supply them with extra cubes or have them borrow cubes for a few minutes from another group.) Tell them to record the number of cubes on the blackline page "Sizing Up Funny Shapes." (This method is a good one to use to estimate the volume of an irregularly shaped object.)

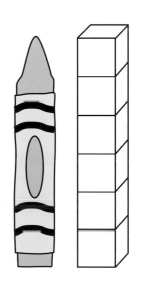

Next, have students partially fill a measuring cup or graduated cylinder with either water or rice. Have students read the level on the container and record it on the blackline page. Then tell them to place the object that they are measuring carefully into the measuring cup or graduated cylinder. If they are using rice instead of water, coach them to push the object firmly into the rice until it is covered. Have them read the new level in their measuring tool and record the number of milliliters displaced by the object.

Students should now compare the number of cubes that they used to replicate the object and the number of milliliters that the object displaced. Have students share their findings. Emphasize that measuring volume by using the cubic centimeters gives an *estimate* of the volume of the irregularly shaped object.

Extend

Ask your students to find the volume of an item that is too big to fit in their measuring cups or graduated cylinders. Have the students design a process and use it to make the measurement.

For example, one method that they might propose would be to find a very large container, fill it to the brim with water, and submerge the large object in the water, being careful to collect all the water that spilled out. Then they could use their measuring tools (calibrated measuring cups or graduated cylinders) to measure the overflow.

Another method that students might suggest would also have them pouring water into a large container, but without filling it all the way to the top this time. Then they could mark the water level before submerging the object that they wanted to measure. They could submerge their object, mark the new level of the water, and then, with their object still submerged in the container, they could begin pouring off water, collecting it all carefully, and continuing until the water remaining in the container had fallen to its original level. Provided that their object was still fully submerged, they could find its volume by using their measuring tools to measure the amount of water that they had poured out. The number of milliliters of water poured out would be equal to the volume of their object in cubic centimeters.

Assessment

To assess your students' understanding of their work, observe them closely as they measure their irregularly shaped objects first by replicating them with one-centimeter cubes and then by using displacement. You should note how successful the students are in replicating the object and ask them how they can decide when they have used enough centimeter cubes. Do students frequently overestimate? Or do they usually underestimate? How accurately do students "read" the measuring cups and graduated cylinders? Do they try to read them from above or below, or do they read them appropriately, at eye-level?

You could also assess your students' understanding of displacement as a measuring strategy through an individual journal assignment. Ask the students to describe the processes that they used to estimate the volume of a particular item and then to determine its actual measurement. Tell them to be sure to explain the difference in the measurements that they obtained by using the two methods.

Where to Go Next in Instruction

Students must call on their problem-solving skills when they are asked to devise their own methods for measuring volume. You can set up a learning center in your classroom to allow your students to measure the volumes of other irregularly shaped objects. The next activity, Weighing In at the Carnival, uses learning centers to give students opportunities to apply customary and metric units of weight.

Weighing In at the Carnival

Grades 4–5

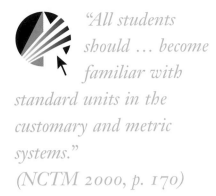

Goals

- Estimate the weight of one kilogram
- Identify items that weigh about one kilogram
- Estimate the weight of one pound
- Compare the relative sizes of standard units of weight in the customary and metric systems

Prior Knowledge

Students should be familiar with the terms *kilogram* and *pound* and have developed benchmarks for both. Students should also have had opportunities to become skillful in estimating the size of items.

Materials and Equipment

pp. 141–42

- A copy of the blackline master "Carnival Recording Sheet" for each team of three or four students
- A standard soccer ball (size 5)
- A standard baseball bat
- A balance scale marked in both pounds and kilograms
- A bathroom scale, calibrated in both pounds and kilograms, if possible
- A small handbell or whistle
- A copy of the template "Award Ribbons" (on the CD-ROM)
- Three identical clear plastic containers
- Vegetable oil, water, and lentils (to fill the three containers to the same level)

To set up five "carnival booths"—

- Booth labels (templates on the CD-ROM)

For booth 1 (Pack a Kilogram) and booth 2 (Just One!)—

- Two heavy bags or cloth sacks (one for each booth)
- A variety of items of different sizes and weights, such as a chalkboard eraser, assorted books, a coffee mug, a teakettle, assorted pencils, assorted notebooks, a sneaker, a pair of scissors, a globe, a soccer ball, a basketball, a sweater, an iron, a potted plant, apples, potatoes

For booth 3 (Make a Kilogram)—

- Modeling clay (allow two to three kilograms per team of three or four students)

This activity has been adapted in part and with permission from Tierney et al. (1998), pp. 53, 56, 57, 62, and 63.

See the accompanying CD-ROM for templates for booth labels that include instructions to students.

For booth 4 (Grocery Lineup)—

- Six grocery items with labels showing different weights (including at least one item with metric units only, at least one item with customary units only, and at least one item with units in both systems)

For booth 5 (Pick Out a Pound)—

- One-pound quantities of five different items (such as coffee, dried soup, unpopped popcorn, pasta, pennies, candy, rocks, sand) in identical, unmarked containers (such as one-gallon sealable clear plastic bags)

Classroom Environment

Working in teams of three or four, students spend approximately two one-hour class periods rotating among five learning centers set up as carnival booths.

Activity

Engage

Technically, weight and mass are distinct quantities, as noted in chapter 2. Mass is the amount of matter in an object, and weight is the measure of the pull of gravity on the object. Most elementary school students are aware that space travel can overcome the pull of Earth's gravity and that objects can become weightless in space. However, students are unlikely to know that the mass of an object remains constant in space travel even though its weight may change. Since most of us will probably remain on Earth, we often blur the distinction between weight and mass, and it may be completely lost on students in grades 3–5. You might want to make the distinction in a discussion with your students, but it is not important to dwell or insist on it at this level.

Focus your students' attention on weight by asking them to name standard units for weight in both the metric and customary systems. Make a list of the units that they identify, such as ounce, pound, ton, milligram, gram, and kilogram. Say, "People in many parts of the world weigh themselves and the objects around them in kilograms. People in some other places usually weigh themselves and the objects around them in pounds." Ask the students, "Which is heavier—a pound or a kilogram?"

Some students will probably need benchmarks for one pound and one kilogram to help them answer the question. Pass around a regulation soccer ball (size 5), which weighs approximately one pound, and a regulation baseball bat, which weighs approximately one kilogram. Let the students handle the objects and familiarize themselves with the weights.

Then call on a student to show the class his or her backpack. Have the student put into the backpack everything that he or she carried to school that day. Ask your students to estimate the weight of the backpack. Allow the students to pick up the backpack before making their estimates. Have half the class give estimates in pounds and the other half give estimates in kilograms. Write the students' estimates on the board.

Then use a bathroom scale to weigh the backpack. (If you are unable to obtain a scale that is calibrated in both pounds and kilograms, use 1 kg ≈ 2.2 lbs to convert from one system to the other.) Weigh the backpack in both metric and customary units, and identify the estimates—

one in pounds and one in kilograms—that are closest to the actual weight of the backpack. Tell your students that they may want to think about the weight of the backpack (as well as the weights of the soccer ball and the baseball bat) as they work through the next part of the activity. (You might even leave these items out on a desk for students to refer to or handle as they work.)

Explore

Say to your students, "You're going to take part in a mathematics carnival right here in our classroom. Our carnival has the special theme 'Weighing In,' and you'll be working in a team to visit five carnival booths."

Show the students the five learning centers that you've set up in your classroom, and point to the posters (from the templates on the CD) that name each center as one of the carnival booths: Pack a Kilogram (booth 1), Just One! (booth 2), Make a Kilogram (booth 3), Grocery Lineup (booth 4), and Pick Out a Pound (booth 5).

Assign the students to teams of three or four, and give each team a number. Tell the students that they will be moving from one carnival booth to another, and at each booth they will find materials that they will use to perform a task that in some way involves estimating a pound or a kilogram. Be sure to say that they will not be using a scale until the "weighing in" at the end of the carnival!

Explain that you will be the carnival director, and you will tell the teams at which booth to start. You will also signal when they are to move to the next booth. Ring a bell (or blow a whistle or clap your hands) to demonstrate the signal that you've chosen for this purpose during the carnival.

Distribute a copy of the blackline master "Carnival Recording Sheet" to each team. Explain that one team member must serve as a recorder, and this student will keep a record of the team's work at each booth. (You may want to select the recorder for each booth activity.) Tell the students that they will be turning the sheets in to the carnival director when all the teams have visited all the booths.

Explain to the students what they will be doing at each booth. Go to each one in turn, and show the students the materials located at each:

- At booth 1 (Pack a Kilogram), the teams find a bag that they fill with items available at the booth until they estimate that the bag weighs one kilogram.
- At booth 2 (Just One!), the teams find another bag that they fill with just one object that they estimate weighs one kilogram.
- At booth 3 (Make a Kilogram), the teams find modeling clay that they use to make a ball that they estimate weighs one kilogram. (They will carve their team's number on their ball and take it with them to turn in to the carnival director along with their completed recording sheet.)
- At booth 4 (Grocery Lineup), the teams use estimates to arrange an assortment of grocery products from lightest to heaviest. (The items' labels indicate their weights, but some are in metric units and some are in customary units.)
- At booth 5 (Pick Out a Pound), the teams select an item that they estimate weighs one pound.

Use the template "Award Ribbons" on the CD to make enough awards for some teams to tie as winners at some booths and for all the teams to be winners at booth 5.

(This last booth is designed to help students discover that items that weigh a pound can look very different. The booth offers a selection of containers, each of which holds a one-pound quantity of a particular material, but the volumes of the materials differ.)

To begin the investigation, send each team to a starting booth. Emphasize that each team member should have an opportunity to hold any object that the team is considering so that everyone can give an opinion about its weight. Remind the teams that they will work at the first booth until you ring the bell (or give whatever signal you prefer) to send each team to the next booth (or from booth 5 to booth 1). Allow students to work at each booth for about ten to fifteen minutes. Explain that the teams will continue to move from booth to booth until they have visited them all. For additional atmosphere, you could play music with a carnival theme each time you signal the end of a session at a booth.

Listen to the mathematical conversations that develop as the teams work at the booths, and, if necessary, encourage students to use their benchmarks. See if students are making appropriate comparisons and contrasts. Listen for comments such as "I have objects in both of my hands, and the one in my left hand feels heavier," or "This feels as heavy as the baseball bat, so I think it weighs about one kilogram." Also take note of any misconceptions that students have. For instance, you might hear a student say, "This box is bigger so it must weigh more." Students' comments as they work in booth 5 (Pick Out a Pound) may be especially interesting. You may have some students who actually come to the correct conclusion that the items at this booth are all the same weight.

As you move around the room, make sure that the teams are giving all members a chance to hold each item and offer opinions about its weight. If necessary, remind them about appropriate behavior for cooperative work in groups.

At the end of the carnival, when all the teams have visited all the booths, have each team hand its sheet in to you in your role as carnival director. Then have the class help you determine which team is the "winner" at each booth. To do this, use a balance scale or the bathroom scale that you used earlier to weigh the backpack, and "weigh in" everything named by the teams on their sheets. (You might also borrow the scale from the school nurse's room to help with this measuring.)

To weigh in at booths 1 and 2 (Pack a Kilogram and Just One!), fill the bag again and again according to the items listed on the teams' record sheets and weigh the bag each time on a scale. Ask the members of each team how they used their benchmarks to help make their decisions.

Booth 3 (Make a Kilogram) offers an additional opportunity for ordering and estimating. Display the clay balls made by all the teams, and ask students to heft them and order them from lightest to heaviest. (If they think two balls have the same weight, they should line them up one behind the other.) Then ask the class to predict the winner—that is, the one closest to one kilogram in weight. Finally, weigh each ball to determine the winner by actual measurements.

By working at booth 4 (Grocery Lineup), students should be able to estimate how many grams equal one pound (454 gm ≈ 1 lb.) and how many grams equal one ounce (28 gm ≈ 1 oz.). Their work will help

them relate the metric units gram and kilogram with the more familiar customary units ounce and pound.

Students may be surprised that all the teams are winners at booth 5 (Pick Out a Pound). This discovery will provide you with an excellent opportunity to discuss *volume* versus *weight*. Things that look different because they occupy different amounts of space can actually have the same weights.

In identifying the winners at each booth, make sure that you also discuss the mathematics behind the task that the students performed at the booth. Constantly stress the effectiveness of using benchmarks for estimates.

Extend

Encourage students to look again at the items at booth 5 (Pick Out a Pound). Call their attention to the differences in the space occupied by each of the items that weigh one pound. Ask them to order the items from the one that takes up the least space—that is, the item with the least volume—to the one that take up the most space—the item with the greatest volume.

Introduce the term *density*. Explain that density is the mass of matter per unit of volume. Dense things are heavy for their size since they have a lot of matter in a relatively small space.

Fill each of three identical clear plastic containers to the same level with water, olive oil, or lentils. Ask students to write predictions about which container will weigh more, and why. Encourage them to examine and hold each container. Hold a class discussion in which students compare their answers. Finally, use a scale to determine actual weights of the three containers.

Point out that in this activity, volume remains constant while weight varies because of the different densities. Help your students understand that this situation is the opposite of that in booth 5, where weight remains constant while volume varies.

Next, ask your students, "What do you think would happen if we poured some beans into the water? Which would remain on top, the beans or the water? What if we poured some oil into the water?" Have students make predictions and then actually experiment to find the answers to these questions.

Assessment Ideas

Students can assess their own abilities to use benchmarks to estimate weight by comparing their estimates at the booths with the actual weights measured at the end of the carnival. The students' work during the carnival also lends itself nicely to the writing of journal entries, which you can then use for assessment.

To help you assess your students' understanding of the weight of a kilogram, the similarities and differences between kilograms and pounds, and the similarities and differences between volume and weight, you can ask them to write about these topics in their journals. Some possible writing prompts include the following:

- "Suppose that you are on a trip to Spain, and you visit the outdoor fruit and vegetable market in Barcelona. Here metric units are used to measure everything. All items are sold by the kilogram, which means that the price of each is determined by how much of the item or how many such items make up one kilogram. Think about

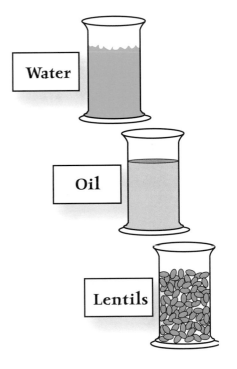

the carnival booths where you determined how much of an item or how many such items weighed one kilogram. Choose a fruit or vegetable that you would want to buy if you were at the Barcelona market. Tell how you would estimate (without a scale) how much of this item or how many such items would weigh one kilogram." (Students might write that they would choose a cantaloupe, for example, and would heft it to compare its weight to that of some benchmark for a kilogram.)

- (Extension for home) "Choose one or more examples of some fruit or vegetable that you find at home. Without using a scale, estimate the weight of your selection. Tell how many people you think you would need to eat your fruit(s) or vegetable(s). Explain your thinking." (A student might write that at the carnival, she found that 2 apples weighed a little less than half a kilogram, so she would choose 4 apples and share them with her brother, mother, and father, with each person in the family having one apple.)

- "Compare a kilogram and a pound. How are they alike? How are they different? List some items that weigh one kilogram and some items that weigh one pound." (Students might write that a kilogram is the standard metric unit for measuring weight, and a pound is the standard customary unit. They might say that a kilogram is about twice as heavy as a pound. They might say that a pair of sneakers weighs about one pound, and an iron weighs about one kilogram.)

Where to Go Next in Instruction

See the CD-ROM for the article "This Little Piggy" (Brahier, Kelly, and Swihart 1999), which provides a hands-on activity on weight, volume, and money.

Continue to engage your students in real-world activities that give them additional experiences in estimating and measuring weight. If possible, take a field trip to a pumpkin patch, an apple orchard, or a farm that grows watermelons, depending on the climate and the time of year. Or arrange an excursion to a local grocery store to examine some particular type of fresh fruit in stock. Arrange for pairs of students to select a sample that they consider to be of "average" weight for that type of fruit. They can estimate the weight of their piece of fruit, and then use an available scale to determine a more exact weight. They can write both numbers—estimated weight and measured weight on the scale—on separate index cards.

Back in the classroom, students can arrange both sets of cards in order from lowest to highest and attach them with clothespins to two "number lines" made with clothesline stretched across the room. Work with your students to determine the median and the mode of each set of values. If your students have mastered division, they can also find the mean of each set.

Ask your students the following questions:

- "Were you good estimators?"
- "How do you know?"
- "How do the number lines help you know if your estimate was a good one?" (Students can look at the placement of their estimate and the placement of their actual weight and can calculate the difference.)
- "How can you become better at estimating?"

Now that students have developed a feel for the weight of an "average" example of a particular type of fruit, have them make a guess of the heaviest one ever recorded. Let them conduct research to find the actual weight.

This chapter has shown how to help students in grades 3–5 extend and connect their earlier experiences in measuring two-dimensional shapes with their discoveries about measuring three-dimensional objects. For example, students explore how to use the area and height of a rectangular solid to determine its volume.

Activities in the chapter also let students use concrete experiences to develop strategies for finding the surface areas of rectangular solids and the volumes of irregular solids. *Principles and Standards for School Mathematics* emphasizes that "these concrete experiences are essential in helping students understand the relationship between the measurement of an object and the succinct formula that produces the measurement" (NCTM 2000, p. 175). Such experiences with measurement in three dimensions in grades 3–5 is essential and provides a springboard for more advanced activities with volume and surface area at the middle school level.

For the heaviest fruits on record, check the *Guinness Book of World Records,* or go to http://waynesword .palomar.edu/pljan96.htm.

NAVIGATING *through* MEASUREMENT

Looking Back and Looking Ahead

Measurement is a lifelong skill. In prekindergarten through grade 2, students should encounter measurement through an informal approach that fosters discovery through exploration. Investigations at this early stage should focus on the lengths, areas, volumes, and weights of particular objects. These first experiences of measurement involve covering, filling, comparing, and matching objects according to specific attributes, and students should use estimates and nonstandard units of measure to make direct and indirect comparisons of objects.

The foundational ideas that students develop in prekindergarten through grade 2 lead to increased understanding and skill in measuring throughout the later grades. In grades 3–5, measurement experiences develop and deepen connections that students have made between measurable attributes, including length, area, volume, angle, weight, time, and temperature, and appropriate units for measuring them. As students expand their ideas about measurement, they also should expand their vocabularies for talking about the topic. They can improve their ability to express what they know by exploring such questions as—

- "What do we mean when we use the word *length*?"
- "What is *volume*?"
- "How is measuring *area* different from measuring *perimeter*?"

The answers to all such questions should be embedded in everyday situations. For example, students can investigate volume by considering how much milk students drink at school each day. They can explore length by examining the distance from their school to a nearby city park.

The focus should always be on measurement as a vehicle to "connect ideas within areas of mathematics and between mathematics and other disciplines" (NCTM 2000, p. 171).

Students in grades 3–5 should continue to estimate measurements and use nonstandard units as a bridge to understanding and using standard units of measure. For example, students in grade 3 may explore the different ways in which a container can be filled with different objects, such as two-centimeter cubes and two-inch cubes, to cement their awareness of the inverse relationship between the number of objects needed to fill a container and the size of the objects used to fill it.

As students in grades 3–5 make the transition from nonstandard to standard units, they should have opportunities to explore units in different measuring systems. When students are weighing an object, they should know that weight is customarily measured in ounces, pounds, tons, grams, kilograms, and so on. They should know that perimeter is a length and is typically expressed in inches, feet, yards, centimeters, meters, kilometers, and so on, and that area is a surface and is measured in the squares of these units. Repeated explorations familiarize them with the length of a meter and the weight of a pound, for example, and these quantities become benchmarks for them to use in future measurement tasks. Their understanding of units grows when they make their own measuring tools, such as rulers and spring balance scales.

Students in grades 3–5 should become skillful in selecting an appropriate standard unit and an appropriate measuring tool for the task at hand. They should know, for example, that a twelve-inch ruler would be challenging to use to measure the length of a school hallway, but a yardstick or a measuring tape would make the task fairly easy to complete. They should also recognize the value of fractional units in achieving precision and know that context and purpose dictate how precise a measurement needs to be.

Students in grades 3–5 should investigate the areas of rectangles, parallelograms, and triangles by probing the relationships among these shapes when they share the same base and height. Their explorations should lead them to develop and use formulas for area.

These central ideas and skills provide a foundation for the concepts and skills that students will develop in the middle grades. Measurement experiences in grades 6–8 should extend and deepen students' understanding. Middle school students continue to select and use compatible units to measure given attributes, estimate measurements, consider units and scales in relation to a desired degree of precision, and solve real-life problems involving the perimeter and area of two-dimensional shapes and the surface area and volume of three-dimensional ones.

It is important to recognize that "measurement is far more complex than we realize" (Van de Walle 2001, p. 277). We should also be aware that it is through the experiences that we orchestrate for our students across the grades that they develop an understanding of measurement and become capable of applying what they know meaningfully and skillfully.

NAVIGATIONS SERIES

GRADES 3–5

NAVIGATING *through* MEASUREMENT

Appendix

Blackline Masters and Solutions

What Can We Measure?

Name _____

A member of your group has selected an object with some of the measurable attributes that your class has discussed.

1. What is your object? _____

2. Draw a picture of your object.

3. Use the chart to list each attribute of your object that you could measure, and tell how you would measure it with some other object or objects in the classroom.

Attribute	How would I measure it?

Measurement Madness

Names _____

_____

Your team is going to visit six different measurement stations in your classroom. Each station will give you a unit for measuring a particular attribute of an object. A different member of your team will act as the recorder at each station. The other team members will select objects to compare with the unit at the station. Each teammate will have a turn to be the recorder, who will check each selection to see how it really compares with each unit. The recorder will circle ✓ if the selection is a good one or ✗ if it is not .

Length Station   Unit of length _____

Team member	Item that is shorter than the unit	circle	Item that is equal to the unit	circle	Item that is longer than the unit	circle
1.		✓ ✗		✓ ✗		✓ ✗
2.		✓ ✗		✓ ✗		✓ ✗
3.		✓ ✗		✓ ✗		✓ ✗
Recorder 4.						

Weight Station   Unit of weight _____

Team member	Item that is lighter than the unit	circle	Item that is equal to the unit	circle	Item that is heavier than the unit	circle
1.		✓ ✗		✓ ✗		✓ ✗
2.		✓ ✗		✓ ✗		✓ ✗
Recorder 3.						
4.		✓ ✗		✓ ✗		✓ ✗

Measurement Madness (continued)

Names _____

_____

Volume Station

Unit of volume _____

Team member	Box with less volume than the unit	circle	Box with about the same volume as the unit	circle	Box with greater volume than the unit	circle
1.		✓ ✗		✓ ✗		✓ ✗
Recorder 2.						
3.		✓ ✗		✓ ✗		✓ ✗
4.		✓ ✗		✓ ✗		✓ ✗

Capacity Station

Unit of capacity _____

Team member	Jar with less capacity than the unit	circle	Jar with about the same capacity as the unit	circle	Jar with greater capacity than the unit	circle
Recorder 1.						
2.		✓ ✗		✓ ✗		✓ ✗
3.		✓ ✗		✓ ✗		✓ ✗
4.		✓ ✗		✓ ✗		✓ ✗

Measurement Madness (continued)

Names _____

_____

Area Station   **Unit of area** _____

Team member	Rectangle whose area is smaller than the unit	circle	Rectangle whose area is about the same as the unit	circle	Rectangle whose area larger than the unit	circle
1.		✓ ✕		✓ ✕		✓ ✕
2.		✓ ✕		✓ ✕		✓ ✕
3.		✓ ✕		✓ ✕		✓ ✕
Recorder 4.						

Time Station   **Unit of time** _____

(At the time station, your team will perform some additional tasks.)

Team member	Task that takes less time than the unit	circle	Task that takes about the same time as the unit	circle	Task that takes more time than the unit	circle
Recorder 1.						
2.		✓ ✕		✓ ✕		✓ ✕
3.		✓ ✕		✓ ✕		✓ ✕
4.		✓ ✕		✓ ✕		✓ ✕

Ants' Picnic

Name _____

Imagine that you and the other members of your group are a team of picnic basket ants, and you have just reached a picnic basket filled with supplies and food. Suppose that you are small, but you walk on long, narrow feet that are the size and shape of paper clips! Your teacher has given you a set of picnic items. Each one of you must check out one of the items by stepping around the item's edge with your unusual feet.

1. Select an item and draw a picture of it.

2. Look at the picnic basket item that you selected, and estimate the number of paper-clip footsteps that you would take to walk around it, heel to toe. Record this estimated number of footsteps on your picture of your picnic basket item.

3. Now place paper clips around the edge of the picnic item to see how many of your unusual ant footsteps you would actually need to take to walk all the way around the item.
   a. How many paper clips did you need altogether? _____
   b. How many did you need for each side? Write the number for each side on the corresponding side in your drawing.

Name _____

4. Select a different picnic item. Draw and label a picture of it.

5. Measure each side of your new picnic item in your paper-clip footsteps, and label the length of each corresponding side in your drawing.

6. By participating in the ants' picnic, have you discovered what the word *perimeter* means? _____ In your own words, tell what you think a perimeter is.

Off to the Hardware Store

Name _____

Your teacher has given you a catalog or flyer from a hardware store or a building supply store. Look at the merchandise that the store sells. Use the chart below to make a record of what you find.

- In column 1, list items that sell by a particular measurement attribute (length, area, perimeter, volume, etc.).
- In column 2, give the unit of measure that the store uses to sell each item.
- In column 3, record the number of the page where you found the item in your flyer or catalog.

Item (sold by length, area, perimeter, volume, etc.)	Unit of measure used to sell the item	Page number in the catalog
Length		
1.		
2.		
3.		
Area		
1.		
2.		
3.		
Perimeter		
1.		
2.		
3.		
Volume		
1.		
2.		
3.		
Capacity		
1.		
2.		
3.		

Continued on next page

Off to the Hardware Store (continued)

Name _____

Item (sold by length, area, perimeter, volume, etc.)	Unit of measure used to sell the item	Page number in the catalog
Weight		
1.		
2.		
3.		
Other Attribute _____		
1.		
2.		
3.		

Measurement Scavenger Hunt

Name _____

You are going on a three-stage measurement scavenger hunt with your group. In the first stage, you'll work together to collect objects of a particular length. In the second stage, you'll find objects of a particular weight. And in the third stage, you'll collect objects with a particular angle.

For each stage of the hunt, your teacher will give your group its most important item of equipment—a standard unit of measure. Your group will also need to take along a transparency of "Measurement Scavenger Hunt—Group Data" and a marking pen so that you can make a record of the group's results.

After each stage of the hunt, your group will share its data with the class by showing the transparency on an overhead projector.

Length

1. What is your group's unit of length? _____

2. Work together to find five objects that you think have that length. List your group's objects in the first column of the length chart.

Length Chart

Five objects compared with _____ (your unit of length)

Object	The same length as the unit	A little longer than the unit	A little shorter than the unit
1.			
2.			
3.			
4.			
5.			

This activity draws on ideas in Van de Walle, John A., *Elementary and Middle School Mathematics: Teaching Developmentally,* 4th ed. (New York: Addison Wesley Longman, 2001), pp. 291–94.

Navigating through Measurement in Grades 3–5

Name _____

3. Measure each object. Put a checkmark (✔) in the appropriate column of the length chart to show whether the object's length is the same as, a little longer than, or a little shorter than your unit.

4. Which object's length is *closest* to your unit of length? _____

Weight

5. What unit of weight does your group have? _____

6. Work together to find five objects in your classroom that you think have that weight. List your group's objects in the first column in the weight chart.

Weight Chart

Five objects compared with _____ (your unit of weight)

Object	The same weight as the unit	A little heavier than the unit	A little lighter than the unit
1.			
2.			
3.			
4.			
5			

7. Weigh each object. Put a checkmark (✔) in the appropriate column in the weight chart to show whether the object's weight is the same as, a little heavier than, or a little lighter than as your unit.

8. Which object's weight is *closest* to your unit of weight? _____

Measurement Scavenger Hunt (continued)

Name _____

Angle

9. What angle measure does your group have? _____

10. Work together to find five objects in your classroom that you think have an angle of that size. List your group's objects in the first column in the angle chart.

Angle Chart

Five objects compared with an angle of _____ (your angle measure)

Object	Has an angle that is the same size as the given angle	Has an angle that is a little larger than the given angle	Has an angle that is a little smaller than the given angle
1.			
2.			
3.			
4.			
5.			

11. Measure the angle that you're thinking about in each object. Put a checkmark (✔) in the appropriate column in the angle chart to show whether the angle is the same as, a little large than, or a little smaller than your angle measure.

12. Which object has an angle that was closest to your angle measure? _____

Measurement Scavenger Hunt–Group Data

Names _____

_____

Length Chart

Our five objects compared with _____ (our unit of length)

Object	The same length as the unit	A little longer than the unit	A little shorter than the unit
1 _____			
2 _____			
3. _____			
4. _____			
5. _____			

Our object whose weight was *closest* to our unit of length was _____.

This activity draws on ideas from Van de Walle, John A., *Elementary and Middle School Mathematics: Teaching Developmentally*, 4th ed. (New York: Addison Wesley Longman, 2001), pp. 291–94.

Measurement Scavenger Hunt–Group Data
(continued)

Names _____

_____

Weight Chart

Our five objects compared with _____ (our unit of weight)

Object	The same weight as the unit	A little heavier than the unit	A little lighter than the unit
1			
2			
3.			
4.			
5.			

Our object whose weight was *closest* to our unit of weight was _____.

Measurement Scavenger Hunt–Group Data
(continued)

Names _____

_____

Angle Chart

Our five objects compared with an angle of _____ (your angle measure)

Object	Has an angle that is the same size as the given angle	Has an angle that is a little larger than the given angle	Has an angle that is a little smaller than the given angle
1.			
2.			
3.			
4.			
5.			

Our object with an angle that was *closest* to our angle measure was _____.

My Benchmarks

Name _____

Your teacher has posted a list of objects for you to use to explore *benchmarks* in measurement. Choose five objects from the list. For each object, you will select a measurable attribute, such as *length*, *weight*, or *volume*. Then you will go through the following steps:

a. Decide on an appropriate standard unit of measure for your attribute.
b. Choose a suitable *benchmark* for your unit.
c. Use your benchmark to estimate the attribute in your object.
d. Use a standard measuring tool to measure the attribute in your object.
e. Explain how you used your benchmark to make your estimate.
f. Tell whether your benchmark was a good one, and why, or why not.

Sometimes you will be working in customary (English) units, such as inches, pounds, and quarts, and sometimes you will be working in metric units, such as centimeters, grams, and liters.

Use the chart to help you organize your work and record your findings. Your teacher will do the example with you in both customary and metric units.

Example　　　　　　　　Object/Attribute

Doorway / Height

Customary system
a. Appropriate standard unit _____　　b. Benchmark _____
c. Estimate _____　　d. Actual measure _____
e. How did you use the benchmark to make your estimate?

f. Was your benchmark a good one to use? _____ Why, or why not?

Metric system
a. Appropriate standard unit _____　　b. Benchmark _____
c. Estimate _____　　d. Actual measure _____
e. How did you use the benchmark to make your estimate?

f. Was your benchmark a good one to use? _____ Why, or why not?

This activity draws on ideas from Van de Walle, John A., *Elementary and Middle School Mathematics: Teaching Developmentally,* 4th ed. (New York: Addison Wesley Longman, 2001), pp. 292–93.

My Benchmarks (continued)

Name _____

1. Object/Attribute

_____ / _____

Metric system

a. Standard unit _____

b. Benchmark _____

c. Estimate _____

d. Actual measure _____

e. How did you use the benchmark to make your estimate?

f. Was your benchmark a good one to use? _____ Why, or why not?

---

2. Object/Attribute

_____ / _____

Customary system

a. Standard unit _____

b. Benchmark _____

c. Estimate _____

d. Actual measure _____

e. How did you use the benchmark to make your estimate?

f. Was your benchmark a good one to use? _____ Why, or why not?

Name _____

---

3. Object/Attribute

_____/_____

Metric system

a. Standard unit _____   *b.* Benchmark _____

c. Estimate _____   *d.* Actual measure _____

e. How did you use the benchmark to make your estimate?

f. Was your benchmark a good one to use? _____ Why, or why not?

---

4. Object/Attribute

_____/_____

Customary system

a. Standard unit _____   *b.* Benchmark _____

c. Estimate _____   *d.* Actual measure _____

e. How did you use the benchmark to make your estimate?

f. Was your benchmark a good one to use? _____ Why, or why not?

Name _____

5. Object/Attribute

_____/_____

Customary system

a. Standard unit _____   b. Benchmark _____

c. Estimate _____   d. Actual measure _____

e. How did you use the benchmark to make your estimate?

f. Was your benchmark a good one to use? _____ Why, or why not?

My Measurement Activities

Name _____

In our everyday activities, we often make measurements without even thinking about them. To see what measurements you make in a week, keep a list of all the activities that you do that involve measurement.

My Measurement Journal

Time \\ Day	Monday	Tuesday	Wednesday
Before School			
During School			
After School			
Evening			

My Measurement Activities (continued)

Name _____

Continue your journal for a whole week.

My Measurement Journal *(continued)*

Thursday	Friday	Saturday	Sunday

For which activities was it important to have precise measurements?_____

_____

Why?

Conversion Sense

Name _____

Sometimes when we are measuring, we discover that it will help us if we switch from larger units to an equivalent number of smaller units. Or we discover that it will help us if we switch from smaller units to an equivalent number of larger units. Consider the following table of conversions from one unit of time to another:

Table of Time Conversions

Unit	Equivalent Measure
1 day	24 hours
1 hour	60 minutes
1 minute	60 seconds

1. How much time would it take to count one million one-dollar bills? Explain your solution in detail.

2. How many times does your heart beat in half an hour?
   a. Work with your partner to develop a strategy to answer this question.

   b. Compute the answer, and be ready to explain your work in detail.

This activity draws on ideas from Reys, Robert E., Marilyn N. Suydam, Mary M. Lindquist, and Nancy L. Smith, *Helping Children Learn Mathematics,* 7th ed. (Boston: Allyn & Bacon, 2004), p. 410.

Navigating through Measurement in Grades 3–5

Big Cover-Up

Name _____

Your teacher has given you collections of identical triangles, rhombi, squares, and index cards for you to use in measuring area. In addition, your class has made a list of objects whose areas you can measure.

1. *a.* Choose two objects from the list.
   b. Find their areas by using one of the units that you have been given and counting the number of these units that you need to cover the areas.
   c. After you have measured the areas in this way with one unit, do the same thing with all the other units that you have.
   d. Record your findings in the chart.

Object	Number of triangles needed to cover the area	Number of rhombi needed to cover the area	Number of squares needed to cover the area	Number of index cards needed to cover the area
1.				
2.				

2. What relationship did you notice between the size of each unit and the total number of units that you needed to cover the object?

3. Are your measurements fairly exact, or are they useful mainly as estimates? How do you know?

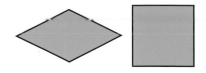

Big Cover-Up Goes Home

Name _____

At home, find three objects that have two-dimensional shapes whose areas you can measure.

1. In columns 1 and 2 of the chart, list each object and a two-dimensional shape that it has.

Areas of Two-Dimensional Shapes in Three Objects at My House

Object	2-D shape	Appropriate square unit	Area measurement
a.			
b.			
c.			

2. What standard square unit (customary or metric) would be appropriate for you to use to measure the area of each shape? Write the units in column 3 of the chart.

3. Measure the area of each two-dimensional shape. You can use paper to make a square unit to use in measuring, or you can measure your shapes without covering them if you prefer. Write your measurements in column 4.

4. Why did you select the unit that you did for each of your objects?

Stuck on Stickers

Name _____

Imagine that you and your partner own the Stuck-on-Stickers Sticker Factory. Business is booming, and the two of you are designing some new sheets of stickers. The stores that want your stickers have certain requirements:

- All your sticker sheets must be rectangular, with dimensions that are whole numbers of inches—no fractions.
- All your stickers must be arranged on the sheets in rows and columns, with no space between them.
- Each of your stickers must take up exactly 1 square inch of space.

1. Suppose that a worker in your factory hands you an order from one of the stores that sells your stickers. This is order no. 1. It gives dimensions of sticker sheets that the store wants. How many stickers will fit on each sheet?

Order No. 1

Item	Length	Width	Number of Stickers
a. Sticker sheet	3 inches	4 inches	
b. Sticker sheet	2 inches	7 inches	
c. Sticker sheet	3 inches	3 inches	

2. Another worker hands you a different order from another store. This is order no. 2. It gives the numbers of stickers per sheet that the store wants. What lengths and widths will sheets have if they contain this many stickers?

Order No. 2

Item	Length	Width	Number of Stickers
a. Sticker sheet			25
b. Sticker sheet			18
c. Sticker sheet			24

Name _____

3. The worker who gave you order no. 2 thinks that your factory could actually come up with more than one rectangular sheet for some of these numbers of stickers. Is she right?_____

4. Use words, pictures, or numbers to describe how you found the total number of stickers that cover sheets *a*, *b*, and *c* in order no. 1.

5. Use words, pictures, or numbers to describe how you decided on the dimensions of sticker sheets *a*, *b*, and *c* in order no. 2.

Stuck on Stickers (continued)

Name _____

6. *a.* What is the biggest sheet that your company would need to fill the orders?_____

b. What is the smallest sheet that your company would need? _____

c. What five sheets would you tell your workers in the Stuck-on-Stickers Sticker Factory to manufacture? Give their dimensions and the number of stickers that each of them would hold.

	Dimensions	Number of stickers
Sheet 1	_____	_____
Sheet 2	_____	_____
Sheet 3	_____	_____
Sheet 4	_____	_____
Sheet 5	_____	_____

Changing Garden

Name _____

Imagine that you are designing a rectangular garden. Also suppose that—

- you want to enclose your garden with 30 feet of wire fencing so that animals won't eat your plants;

- you want to use all 30 feet of your fencing, without any gaps or overlaps; and

- your fencing comes in one-foot sections that you can't split into fractional parts.

Determine all the different ways in which you can arrange the 30 feet of fencing around your garden. Be sure to list the length, width, perimeter, and area of each arrangement.

1. Complete the chart.

Fence	Length (feet)	Width (feet)	Perimeter (feet)	Area (square feet)
A				
B				
C				
D				
E				
F				
G				

2. What does *area* mean?

3. What does *perimeter* mean?

This activity is loosely based on ideas in Lappan, Glenda, James T. Fey, William M. Fitzgerald, Susan N. Friel, and Elizabeth Difanis Phillips, *Covering and Surrounding: Two-Dimensional Measurement*, Connected Mathematics Program (Palo Alto, Calif.: Dale Seymour, 1998), pp. 35–36.

Name _____

4. In the space below, use words, pictures, or numbers to describe how you would find the perimeter of a rectangle of any size.

5. Do all the possible gardens that use the 30 feet of wire fencing have the same perimeter? Why, or why not?

6. Do all the possible gardens that use the 30 feet of wire fencing have the same area? Why, or why not?

Geo-Exploration–Parallelograms

Name _____

You know how to find the area of any rectangle. How could you find the area of any parallelogram? A geoboard, a few rubber bands, and your mathematical thinking skills are really all you'll need to find out. A sheet of geodot paper, a pencil, and this activity page will help, too.

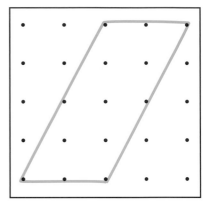

1. Create a parallelogram on your geoboard by stretching a rubber band over four appropriate pegs. Make five different parallelograms on your geoboard in this way.

2. Draw your parallelograms on your geodot paper, and number them from 1 to 5.

3. Use the chart to record the *base* and *height* of each of your parallelograms. Remember that *height* is different from *side*.

Parallelogram	Base	Height	Area
1			
2			
3			
4			
5			

4. Count the squares covered by each parallelogram on your geoboard to find its area. Be sure to add whole squares and parts of squares. Record the areas of your parallelograms in the last column of your chart.

5. Look for patterns in your data. In the space below, explain how you would find the area of *any* parallelogram. Use words, pictures, or numbers to help explain your method.

6. Why do you think your method works? *Hint*: Think about the relationship between rectangles and parallelograms.

_____

This activity is loosely based on ideas in Barson, Alan, "Geoboard Activity Cards–Intermediate" (Fort Collins, Colo.: Scott Resources, 1971).

Navigating through Measurement in Grades 3–5

Geo-Exploration–Triangles

Name _____

Imagine that you want to create a triangular banner that is similar to, but much larger than, the isosceles triangle shown here on a geoboard. The shortest side of your banner is 2 feet, and the distance from the middle of this side to the opposite tip is 4 feet. Can you calculate the area of the banner? To help you do the job, you'll create an assortment of triangles on a geoboard and find their areas. In the process, you'll discover a method for finding the area of any triangle, including the one for your banner.

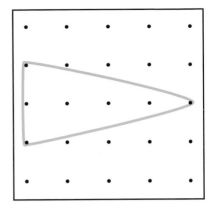

1. Create a triangle on your geoboard by stretching a rubber band over three pegs. Make five different triangles on your geoboard in this way.

2. Draw your triangles on your geodot paper, and number them from 1 to 5.

3. Using the chart shown, record the measures of the base and height of each triangle. Remember that *height* is different from *side*.

Triangle	Base	Height	Area
1			
2			
3			
4			
5			

4. Count the squares covered by each triangle on the geoboard to find its area. Be sure to add whole squares and parts of squares. Record the areas of your triangles in the last column of your chart.

5. Look for patterns in your data. Explain how you would find the area of *any* triangle. Use words, pictures, or numbers to help explain your method.

6. Why do you think your method works? *Hint:* Think about the relationships among rectangles, parallelograms, and triangles.

7. What is the area of the banner?

This activity is loosely based on ideas in Barson, Alan, "Geoboard Activity Cards–Intermediate" (Fort Collins, Colo.: Scott Resources, 1971).

Building Boxes—A

Name _____

Suppose that you and your partner are package designers for the Box-M-Up Package Company. Your department designs boxes without lids, and their dimensions are always whole numbers of units. Box-M-Up wants you to produce scale models of all the boxes that you can create by starting with a rectangle that is 9 units by 11 units. Use grid paper to make models of your boxes, and record their dimensions in the chart below.

	Length (units)	Width (units)	Height (number of layers of unit cubes)	Volume (number of unit cubes needed to fill box)	Total number of unit squares cut from rectangle
Trimmed grid-paper rectangle					
Box 1					
Box 2					
Box 3					
Box 4					
Box 5					

1. What happens to the length, width, and volume as you make each new box?

Box 1 _____

Box 2 _____

Box 3 _____

Box 4 _____

Box 5 _____

2. How did you figure out how many cubes would fit into a box?

Building Boxes–B

Name _____

Suppose that you and your partner are package designers for the Box-M-Up Package Company. Your department designs boxes without lids, and their dimensions are always whole numbers of units. Box-M-Up wants you to produce scale models of all the boxes that you can create by starting with a rectangle that is 9 units by 11 units. Use grid paper to make models of your boxes, and record their dimensions in the chart below.

	Length (units)	Width (units)	Height (number of layers of unit cubes)	Volume (number of unit cubes needed to fill box)	Total number of unit squares cut from rectangle	Surface area (number of unit squares in grid-paper net)
Trimmed grid-paper rectangle						
Box 1						
Box 2						
Box 3						
Box 4						
Box 5						

Describe all the patterns that you can identify from your chart.

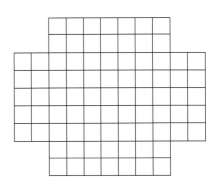

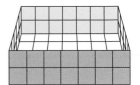

Researching Box Designs

Name _____

Suppose that you and your partner are box designers for the Ooey-Gooey Candy Company. Ooey-Gooey makes caramels and packages them 12 to a box in rectangular boxes that are sold in retail stores. Ooey-Gooey makes boxes with lids. The company wants you and your partner to make all the possible styles of rectangular boxes that will hold exactly 12 caramels, with no space left over. Then the two of you must select one design to recommend for Ooey-Gooey's caramels.

1. Working with grid paper, make a net of each possible box. Cut it out, fold it, and tape the resulting box together.
2. Use the chart to help you organize your research on the boxes. The first row has been filled in as a sample. It shows one possible net. Can you find another?

Dimensions of the box	Volume of the box	Sketch of a net of the box	Surface area of the box
12 x 1 x 1	12 cubic units		50 square units

"Researching Box Designs" is part of the activity Wrap It Up! which has been adapted with permission from Lappan, Glenda, James T. Fey, William M. Fitzgerald, Susan N. Friel, and Elizabeth Difanis Phillips, *Filling and Wrapping: Three Dimensional Measurement*, Connected Mathematics Program (Palo Alto, Calif.: Dale Seymour, 1998), p. 16.

Name _____

3. The Ooey-Gooey Candy Company has asked you and your partner to make a presentation in which you will show a model of the box that you are recommending. You must also explain your reasons for your choice. What will you report to Ooey-Gooey?

4. Use tag board or poster board to make a presentation model of the box that you are recommending, and use markers or other materials to decorate it.

Researching Box Designs– Extension

Name _____

Imagine that the Ooey-Gooey Candy Company has changed its marketing plans and has asked you and your partner to design an all-new, larger box that will hold 18 caramels. Use grid paper and cubes to investigate all the possible box designs, and organize your data in the chart below.

Dimensions of of the box	Volume of the box	Sketch of a net of the box	Surface area of the box

1. What are the dimensions of the 18-caramel box with the greatest surface area?

2. What is the surface area of this box?

3. What are the dimensions of the 18-caramel box with the least surface area?

4. What is the surface area of this box?

"Researching Box Designs–Extension" is part of the activity Wrap It Up! which has been adapted with permission from Lappan, Glenda, James T. Fey, William M. Fitzgerald, Susan N. Friel, and Elizabeth Difanis Phillips, *Filling and Wrapping:Three-Dimensional Measurement*, Connected Mathematics Program (Palo Alto, Calif.: Dale Seymour, 1998), p. 16.

Sizing Up Funny Shapes

Names _____

1. Working as partners, select a "funny-shaped" object from the collection of objects at your displacement station. Identify your object in column 1 of the chart.

Object	Number of one-centimeter cubes in your model (= estimated volume of your object in cc)	Original level of the water (or rice) in your measuring tool (mL)	New level of the water (or rice) in your measuring tool after adding your object (mL)	Number of milliliters of water (or rice) displaced by your object (mL)	Actual volume of your object (mL)
1.					
2.					
3.					
4.					
5.					
6.					
7.					
8.					

This activity has been adapted with permission from the Teaching Integrated Mathematics and Science (TIMS) Project, *Math Trailblazers: Grade 4,* (Dubuque, Iowa: Kendall/Hunt Publishing Company, 1997), pp. 218–20.

Names _____

2. Use the one-centimeter cubes at your station to construct a rough "copy" of your object.

3. Count the cubes in your model, and record the number in column 2 of the chart. *This is your estimate of the volume of your object, in cubic centimeters (cc).*

4. From the assortment of measuring tools at your station, choose a tool that you think will be large enough to contain your object, with some room to spare.

5. Don't put your object into your measuring tool yet. Carefully fill your measuring tool with what you think will be enough water (or rice) to cover your object when you submerge it. Read the level of the water (or rice) on the measuring tool, and record the number in column 3 of the chart. *This is your measurement of the volume of the water (or rice), in milliliters (mL).*

6. Now put your object carefully into the water (or rice). (If you're working with rice, push the object gently but firmly into the rice until it is covered.) Read the new level on your measuring tool. Record this new number in column 4 of your chart. *This is your measurement of the volume of the water (or rice) plus the volume of your object, in milliliters (mL).*

7. Your new level minus your original level is equal to the number of milliliters of water (or rice) displaced by your object. Subtract, and record the difference in column 5 of your chart. *This difference is equal to the actual volume of your object.*

8. Enter the actual volume of your object in column 6 of the chart.

9. Remember that 1 mL = 1 cc. Compare your estimate of your object's volume (in column 2) with your measurement of the actual volume in column 6. Write about how close your estimate was to your actual measurement.

10. Can you rebuild your "copy" of your object to get a closer, more accurate estimate? _____ If so, what will you change?

11. Repeat this entire process to measure the volumes of seven other funny-shaped objects at your station. Record your results in the chart.

Carnival Recording Sheet

Team number _____

Team recorder _____

Team members_____

_____

_____

Your team is about to participate in a mathematics carnival with the special theme "Weighing In." You will visit five carnival booths, and you will use the materials at each one to perform a task that calls on you to estimate a pound or a kilogram. Follow the instructions posted at each booth, and continue to work at each one until the carnival director (your teacher) signals you to move to the next booth.

Your team recorder should use this sheet to keep a record of your work at each booth. At the end of the carnival, your team will hand this sheet in to the carnival director, who will then help you find out if your team has won any blue ribbons.

Booth 1—Pack a Kilogram

List the items that you put into the bag to give it an estimated weight of one kilogram.

_____

_____

_____

_____

Booth 2—Just One!

Identify the single item that you put into the bag to give it an estimated weight of one kilogram.

This activity has been adapted in part with permission from Tierney, Cornelia, Margie Singer, Marlene Kliman, and Megan Murray, *Measurement Benchmarks: Estimating and Measuring, Grade 5 Investigations*, TERC Investigations in Number, Data, and Space (White Plains, N.Y.: Dale Seymour, 1998), pp. 53, 56, 57, 62, and 63.

Booth 3—Make a Kilogram

When your team has completed its ball, carefully carve your team number in it.

Booth 4—Grocery Lineup

List the grocery items in order from lightest to heaviest.

Booth 5—Pick out a Pound

Record the item that your team thought weighed one pound.

Solutions for the Blackline Masters

Students' responses will vary for "What Can We Measure?" "Measurement Madness," "Ants' Picnic," "Off to the Hardware Store," "Measurement Scavenger Hunt," "Measurement Scavenger Hunt"—Group Data, "My Benchmarks," and "My Measurement Activities."

Solutions for "Conversion Sense" are discussed in the text.

Solutions for "Big Cover-Up"

1. Students' answers will vary.
2. The larger the unit, the fewer the units needed to cover the object and measure its area.
3. The measurements are estimates; the units will not cover the objects exactly.

Solutions for "Big Cover-Up Goes Home"

Students' answers will vary.

Solutions for "Stuck on Stickers"

1. In order 1, 12 stickers will fit on sticker sheet (*a*), 14 stickers will fit on sheet (*b*), and 9 stickers will fit on sheet (*c*).
2. Possible lengths and widths of the sticker sheets in order 2 are shown:

Order No. 2

Item	Length	Width	Number of stickers
a. Sticker sheet	1	25	25
	5	5	
b. Sticker sheet	1	18	18
	2	9	
	3	6	
c. Sticker sheet	1	24	24
	2	12	
	3	8	
	4	6	

3. Yes, the worker is correct. Sticker sheet (*a*) could be 1 × 25, 5 × 5, or 25 × 1. Sticker sheet (*b*) could be 1 × 18, 2 × 9, 3 × 6, 6 × 3, 9 × 2, or 18 × 1. Sticker sheet (*c*) could be 1 × 24, 2 × 12, 3 × 8, 4 × 6, 6 × 4, 8 × 3, 12 × 2, or 24 × 1.

4 and 5. Students' responses will vary but should express the idea that the total number of stickers is equal to the product of the length and the width.

6. *a.* Sheet (*a*) in order 2 would be the largest, with 25 stickers

   b. Sheet (*c*) in order 1 would be the smallest, with 9 stickers.

   c. Students' responses will vary.

Solutions for "Changing Garden"

1. Students' completed charts should look like the following:

Fence	Length (feet)	Width (feet)	Perimeter (feet)	Area (square feet)
A	1	14	30	14
B	2	13	30	26
C	3	12	30	36
D	4	11	30	44
E	5	10	30	50
F	6	9	30	54
G	7	8	30	56

2. *Area* is the number of square units that cover a two-dimensional figure or shape.

3. *Perimeter* is the distance around a two-dimensional figure or shape.

4. Students' responses will vary but should suggest the idea that they would (*a*) add the lengths of all four sides, or (*b*) add the length and the width of the rectangle and multiply this sum by two, or (*c*) multiply the length by two and the width by two and add these two products together.

5. Yes, all the gardens have the same perimeter. The changes in the dimensions of each arrangement do not affect the total perimeter since at no point does the total *amount* of fencing change.

6. No, all the gardens do not have the same area. As the dimensions change, square units are either "pulled into" or "pushed out of" the garden, changing its area. As the length and width approach each other in value (i.e., as the garden becomes more nearly square), the space inside the garden, or its area, increases.

Solutions for "Geo-Exploration–Parallelograms"

3 and 4. Students' answers will vary. A sample completed chart follows:

Parallelogram	Base	Height	Area
1	1	1	1
2	1	2	2
3	2	2	4
4	2	3	6
5	3	3	9

5. Students' responses will vary but should suggest the idea that the area of a parallelogram is equal to the product of its base and its height.

6. Students' responses will vary. To explain why the method works, they might draw a parallelogram. Then they might superimpose a drawing of a rectangle with the same base and height. They might reconfigure the area of their parallelogram to show that it is equal to the area of the rectangle. They might then say that they know how to find the rectangle's area—by multiplying its length and width. But this is also the area of parallelogram. And they know that the length and width of the rectangle are equal to the base and height of the parallelogram. So they can find the area of their parallelogram—and of any parallelogram—by multiplying its base and height.

Solutions for "Geo-Exploration–Triangles"

3 and 4. Students' answers will vary. A sample completed chart follows:

Triangle	Base	Height	Area
1	1	1	0.5
2	1	2	1.0
3	1	3	1.5
4	2	3	3.0
5	2	4	4.0

5. Students' responses will vary but should suggest the idea that the area of a triangle is equal to one-half of the product of its base and its height.

6. Students' responses will vary. To explain why the method works, they might draw a triangle. Then they might superimpose a drawing of a parallelogram with the same base and height. They might show that the area of their triangle is one-half the area of the parallelogram. They might then say that they know how to find the parallelogram's area—by multiplying its base and height. But this is two times the area of the triangle. So they can find the area of their triangle—and of any triangle—by multiplying its base times its height times 1/2.

7. The area of the banner is 4 square feet.

Solutions for "Building Boxes"—A and B

1. The completed chart shows the volumes and surface areas of boxes made from an 11-by-9 rectangle:

	Length (units)	Width (units)	Height (number of layers of unit cubes)	Volume (number of unit cubes needed to fill box)	Total number of unit squares cut from rectangle	Surface area (number of unit squares in grid-paper net)
Trimmed grid-paper rectangle	11	9	0	0	0	99
Box 1	9	7	1	63	4	95
Box 2	7	5	2	70	16	83
Box 3	5	3	3	45	36	63
Box 4	3	1	4	12	64	35
Box 5	(no box)					

Patterns are discussed in the text, and table 4.1 shows the volumes and surface areas of boxes made from a 9-by-9 square.

Solutions for "Researching Box Designs"

2. Figure 4.4 in the text shows a sample completed chart.
3. Students' responses will vary.

Solutions for "Researching Box Designs"—Extension

Students' nets will vary. Other answers are discussed in the text.

Solutions for "Sizing Up Funny Shapes and "Carnival Recording Sheet"

Students' answers will vary.

References

Barson, Alan. "Geoboard Activity Cards—Intermediate." Fort Collins, Colo.: Scott Resources, 1971.

Brahier, Daniel J., Monica Kelly, and Jennifer Swihart. "This Little Piggy." *Teaching Children Mathematics* 5 (January 1999): 274–80.

Cathcart, W. George, Yvonne Potheir, James Vance, and Nadine Bezuk. *Learning Mathematics in Elementary and Middle Schools*. Upper Saddle River, N.J.: Prentice Hall, 2001.

Civil, Marta, and Leslie Khan. "Mathematics Instruction Developed from a Garden Theme." *Teaching Children Mathematics* 7 (March 2001): 400–405.

Gavin, M. Katherine, Louise P. Belkin, Ann Marie Spinelli, and Judy St. Marie. *Navigating through Geometry in Grades 3–5. Principles and Standards for School Mathematics* Navigations Series. Reston, Va.: National Council of Teachers of Mathematics, 2001.

Lappan, Glenda, James T. Fey, William M. Fitzgerald, Susan N. Friel, and Elizabeth Difanis Phillips. *Covering and Surrounding: Two-Dimensional Measurement*. Connected Mathematics Program. Palo Alto, Calif.: Dale Seymour, 1998a.

———. *Filling and Wrapping: Three Dimensional Measurement*. Menlo Park, Calif.: Dale Seymour, 1998b.

Lubinski, Cheryl A., and Diane Thiessen. "Exploring Measurement through Literature." *Teaching Children Mathematics* 2 (January 1996): 260–63.

Nitabach, Elizabeth, and Richard Lehrer. "Developing Spatial Sense through Area Measurement." *Teaching Children Mathematics* 2 (April 1996): 473–76.

Pinczes, Eleanor J. *One Hundred Hungry Ants*. Boston: Houghton Mifflin, 1993.

Reynolds, Anne, and Grayson H. Wheatley. "Third-Grade Students Engage in a Playground Measuring Activity." *Teaching Children Mathematics* 4 (November 1997): 166–70.

Reys, Robert E., Marilyn N. Suydam, Mary M. Lindquist, and Nancy L. Smith. *Helping Children Learn Mathematics* (7th ed.). Boston: Allyn & Bacon, 2004.

Teaching Integrated Mathematics and Science (TIMS) Project. *Math Trailblazers: Grade 4*. Dubuque, Iowa: Kendall/Hunt Publishing Company, 1997.

Tierney, Cornelia, Margie Singer, Marlene Kliman, and Megan Murray. *Measurement Benchmarks: Estimating and Measuring, Grade 5 Investigations*. TERC Investigations in Number, Data, and Space. White Plains, N.Y.: Dale Seymour, 1998.

Van de Walle, John A. *Elementary and Middle School Mathematics: Teaching Developmentally*. 4th ed. New York: Addison Wesley Longman, 2001.

Suggested Reading

Burghardt, Becky, and Ginny Heilman. "Water Matters." *Teaching Children Mathematics* 1 (September 1994): 24–25.

Burns, Marilyn. *About Teaching Mathematics: A K–8 Resource*. 2nd ed. Sausalito, Calif.: Math Solutions, 2000.

Fay, Nancy, and Catherine Tsairides. "Metric Mall." *Arithmetic Teacher* 37 (September 1989): 6–11.

Greenes, Carole, Linda Schulman, and Rika Spungin. *ThinkerMath: Developing Number Sense and Arithmetic Skills, Grades 3–4*. Sunnyvale, Calif.: Creative Publications, 1989.

Hatfield, Mary M., Nancy T. Edwards, and Gary Bitter. *Mathematics Methods for Elementary and Middle School Teachers*. 3rd ed. Boston: Allyn & Bacon, 1997.

Hopkins, Martha, Daniel Brahier, and William Speer. "That's Gross (Tonnage)." *Teaching Children Mathematics* 3 (April 1997): 442–45.

Kubota-Zarivnij, Kathy. "How Do You Measure a Dad?" *Teaching Children Mathematics* 6 (December 1999): 260–64.

Long, Betty, and Deborah Crocker. "Adventures with Sir Cumference: Standard Shapes and Nonstandard Units." *Teaching Children Mathematics* 7 (December 2000): 242–45.

Parker, Janet, and Connie Carroll Widmer. "Teaching Mathematics with Technology: Patterns in Measurement." *Arithmetic Teacher* 40 (January 1993): 292–95.

Wickett, Maryann. "Measuring Up with the Principal's New Clothes." *Teaching Children Mathematics* 5 (April 1999): 476–79.